HYDROLOGICAL IMPACTS OF LAND USE CHANGES ON WATER RESOURCES MANAGEMENT AND SOCIO-ECONOMIC DEVELOPMENT OF UPPER EWASO NG'IRO RIVER BASIN IN KENYA

Hydrological Impacts of Land Use Changes on Water Resources Management and Socio-economic Development of Upper Ewaso Ng'iro River Basin in Kenya

DISSERTATION

Submitted in fulfillment of the requirements of
the Board for Doctorates of Delft University of Technology
and of the Academic Board of the UNESCO-IHE Institute for Water Education
for the Degree of DOCTOR
to be defended in public
on Monday, May 1, 2006 at 10:00 hours
in Delft, the Netherlands

by

Stephen Njuguna Ngigi
born in Nyandarua, Kenya

Master of Science in Agricultural Engineering
(*Soil and Water Engineering Option*)
University of Nairobi, Kenya

CRC Press
Taylor & Francis Group
Boca Raton London New York

CRC Press is an imprint of the
Taylor & Francis Group, an **informa** business

CRC Press
Taylor & Francis Group
6000 Broken Sound Parkway NW, Suite 300
Boca Raton, FL 33487-2742

First issued in hardback 2017

ISBN-13: 978-0-4154-0918-6 (pbk)
ISBN-13: 978-1-1384-6868-9 (hbk)

Visit the Taylor & Francis Web site at
http://www.taylorandfrancis.com

and the CRC Press Web site
http://www.crcpress.com

Synopsis

The need to improve food production in semi-arid environments of sub-Saharan Africa (SSA) cannot be overemphasized. Agricultural production is generally low (1 ton ha^{-1}), which is synonymous with a poverty threshold of US$1 day^{-1}. The situation is aggravated by water scarcity where many countries in SSA are below the per capita water security threshold of 1,000m^3 yr^{-1}. Though the situation seems desperate, all is not lost. One of the most promising solutions is upgrading rainfed agriculture through the adoption of rainwater harvesting and management (RHM) systems, which improve water availability for productive purposes. The study assessed land use changes, in particular adoption of RHM systems, resulting from the need to upgrade rainfed agriculture in terms of improving food production and hydrological impacts on river basin water resources management. RHM systems are diverse and range from in-situ moisture conservation, small to large runoff storage systems and flood diversion and spreading (spate irrigation).

The goal of the study was to provide information required in the formulation of integrated water resources management to enhance socio-economic development and sustain ecological balance in water-scarce river basins. The overall objective was to assess the hydrological impacts of land use changes on water resources management and socio-economic development of upper *Ewaso Ng'iro* river basin in Kenya. This was accomplished through field survey of viable RHM systems, agro-hydrological and hydro-economic evaluation of on-farm storage and in-situ RHM systems, and assessment of the impacts flood storage on dry season water abstractions. A conceptual framework was developed as an analytical tool for assessing hydrological impacts of land use changes in upper *Ewaso Ng'iro* river basin. The study was conducted in semi-arid and semi-humid areas where recent land use changes have led to decline in river flows and conflicts among water users. Water abstraction and flood storage analysis were based on *Naro Moru* sub-basin.

The results revealed that there a number of viable RHM systems, which can improve agricultural production in semi-arid environments. However, their effectiveness is limited due to high water losses (30-50%), inadequate storage capacity (25-50% reliability for 30-50m^3 farm ponds), poor water management, high occurrence (60-80%) of intra- seasonal dry spells (10-15 days) and off-seasonal dry spells (20-30 days) and farmers' risk-averseness and financial constraints to invest in new farming systems. Nevertheless, both on-farm storage systems for supplemental irrigation and in-situ systems (i.e. conservation tillage) were found to be economically viable for smallholder farmers. Economic analysis indicated that a farmer can recover investment cost within 2-3 years.

The assessment of hydrological impacts of RHM systems revealed that they retain 20-30% of runoff, which due to their small acreage, is negligible at a river basin scale. The area under agricultural land is less than 20% in the entire river basin, and the adoption rate of RHM systems is 10-15%. Hence, up-scaling small-scale RHM systems may not significantly reduce river flows. However, flood storage can reduce dry season irrigation water abstractions by more than 50% without affecting hydro-ecological functions downstream. Although the area under irrigation is low (< 5%), it accounts for more than 95% of water abstractions. Low river flow has forced large-scale horticultural farmers to construct reservoirs for flood storage. Thus, there is need to formulate sustainable strategies to ensure equity distribution of basin water. However, detailed hydrological modelling would suffice to investigate future scenarios and provide conclusive results.

Acknowledgements

The study was accomplished through direct and indirect contributions of many persons and institutions. I will start with Dr. Herbert G. Blank (former IWMI Regional Advisor), Dr. Francis N. Gichuki (IWMI, Colombo/University of Nairobi) and Dr. Johan Rockström (Director of SEI) for their ideas during the study's formative stage. Dr. Blank also guided me on search for scholarships. I am thankful to WOTRO, IWMI, the USAID-funded GHARP project and the University of Nairobi for their financial support. I am sincerely grateful to my promoter, Prof. Hubert H.G. Savenije, whose technical guidance, constructive discussions, invaluable contributions and encouragement shaped the outcome of the study. Despite the distance, I felt like Huub was always there with me throughout the journey. I hope I will be able to keep some of his skills, especially on technicalities on journal papers. Dr. Rockström also provided technical support during site selection and experimental set up.

The contributions of my other supervisors, Dr. Johan Rockström, Dr. Gichuki and Dr. Frits W.T. Penning de Vries (formerly at IWMI, South Africa) were outstanding. I also had two mentors, Mr. Daniel Schotanus (UNESCO-IHE) and Prof. Charles K.K. Gachene (University of Nairobi), who provided incredible logistic support. I also thank Dr. C.T. Hoanh (IWMI, Colombo) for his contribution in formulating the conceptual framework. The study was on a sandwich basis and different components were done in different countries. This was made possible by the following persons who made all the necessary arrangements: Ms. Vardana Sharma and Ms. Jolanda Boots (UNESCO-IHE, The Netherlands); Ms. Shanthi Weerasekera and Ms. Thushari Samarasekera (IWMI, Colombo); Ms. Mary Njonge (IWMI, South Africa) and Ms. Martha Hondo (WaterNet, Zimbabwe). Ms. Monicah Gammimba (acting IWMI Administrative Assistant, Nairobi) has been very useful in handling my IWMI's related financial matters. Dr. Douglas Merrey (IWMI Africa Director) has also been very prompt in financial management. My supervisors spared no opportunity to meet and discuss my work, even when they were on transit or attending other business in Nairobi.

The field work was done through the support of two dedicated research assistants (Ms. Josephine N. Thome and Mr. Samuel N. Wainaina) who braved heavy storms collecting data for the study. Ms. Thome also assisted in preliminary data analysis and proof reading of some of the journal papers. Secondary data was obtained from the Kenya Meteorological Department, Ministry of Agriculture and Natural Resources Monitoring, Modelling and Management (NRM) Project. I am also grateful to the farmers who provided land for research sites and socio-economic information. The contributions of the GHARP case studies collaborators are also appreciated. Component of the study cannot. The logistic support provided by the KRA/GHARP Secretariat through diligent services of Ms. Susan W. Kung'u (Administrative Secretary) enhanced communication and collaboration among supervisors and journal editors. Her effort and support were remarkable.

Last, but not least, I would like to acknowledge the support, encouragement, patience and goodwill of my family. Joy, my eight years old daughter, could not comprehend what her dad was doing as a student! My mother, brothers and sisters have been very supportive throughout my academic life. For them all I say a big thank you. I also acknowledge any other person or institution that in one way or the other made this study a success. God bless you all for your contributions, undivided attention and support.

Table of Contents

Abbreviations and Acronyms

ADF	-	African Development Fund
ASAL	-	Arid and Semi Arid Lands
CGIAR	-	Consultative Group on International Agricultural Research
CSE	-	Centre for Science and Environment
CWR	-	Crop Water Requirement
DFR	-	Downstream Flow Requirement
DSS	-	Decision Support System
ESA	-	Eastern and Southern Africa
FDC	-	Flow Duration Curve
FAO	-	Food and Agriculture Organization
GDP	-	Gross Domestic Product
GHA	-	Greater Horn of Africa
GHARP	-	Greater Horn of Africa Rainwater Partnership
GoK	-	Government of Kenya
HASR	-	Hydrological Assessment of up-Scaling RHM
HELP	-	Hydrology for the Environment, Life and Policy
IAHS	-	International Association of Hydrological Sciences
IHE	-	Institute for Water Education
IRIN	-	Integrated Regional Information Networks
ISDS	-	Intra-Seasonal Dry Spell
IWRM	-	Integrated Water Resources Management
IWMI	-	International Water Management Institute
JRRS	-	Joint Relief and Rehabilitation Services
KENDAT	-	Kenya Network for Animal Draught Technology
KCTI	-	Kenya Conservation Tillage Initiative
KRA	-	Kenya Rainwater Association
LEISA	-	Low External Input Sustainable Agriculture
NGO	-	Non Governmental Organization
NRM	-	Natural Resources, Monitoring, Modelling and Management
ODS	-	Off-season Dry Spell
RELMA	-	Regional Land Management Unit
RHM	-	Rainwater Harvesting and Management
RGS	-	River Gauging Station
RWH	-	Rainwater Harvesting
SADC	-	Southern Africa Development Cooperation
SARDEP	-	Semi Arid Rural Development Program
SASE	-	Semi Arid Savannah Environments
SEI	-	Stockholm Environment Institute
SIR	-	Supplemental Irrigation Requirement
SIWI	-	Stockholm International Water Institute
SSA	-	Sub-Saharan Africa
SWAT	-	Soil Water Assessment Tool
SWIM	-	System-Wide Initiative on Water Management
UNDP	-	United Nations Development Program
UNESCO	-	United Nations Educational, Scientific and Cultural Organization
USAID	-	United States Aid for International Development
USDA	-	United States Department for Agriculture
USLE	-	Universal Soil Loss Equation
WMO	-	World Meteorological Organization
WOTRO	-	The Netherlands Foundation for the Advancement of Tropical Research
WRAP	-	Water Resources Assessment Program

Notations and Symbols

Notation/Symbol	Description	Dimension
A	Surface area	L
A_R	Area of river basin	L
A_S	Area under RHM system	L
D_{rz}	Crop rooting depth	L
E_c	Crop water requirement	LT^{-1}
E_m	Sublimation	LT^{-1}
E_o	Potential evaporation	LT^{-1}
E_{pan}	Pan evaporation	LT^{-1}
E_s	Soil evaporation	LT^{-1}
E_w	Open water evaporation	LT^{-1}
h	Water depth	L
K_c	Crop factor	-
I_c	Canopy interception	LT^{-1}
K_y	Yield response factor	-
K_p	Evaporation pan co-efficient	-
n	Side slope	-
N	Sample size	-
P	Precipitation/rainfall	LT^{-1}
$P_{b(P)}$	Probability of exceeding P	-
P_D	Threshold rainfall	LT^{-1}
P_{di}	Drought index	-
P_e	Effective rainfall	LT^{-1}
P_s	Snow fall	LT^{-1}
P_{sd}	Standard deviation of rainfall	LT^{-1}
Q_d	Percolation/seepage	L^3T^{-1}
Q_{dp}	Deep percolation	L^3T^{-1}
Q_g	Groundwater flow	L^3T^{-1}
Q_i	Irrigation water requirement	LT^{-1}, L^3T^{-1}
Q_r	Surface runoff	LT^{-1}, L^3T^{-1}
Q_m	Snow melt flow	L^3T^{-1}
$Q_{r(IS)}$	Surface runoff from in-situ RHM system	L^3T^{-1}
$Q_{r(SS)}$	Surface runoff from storage RHM system	L^3T^{-1}
Q_s	River flow	L^3T^{-1}
q_r	Runoff flow per unit area	L
R_S	System reliability (%)	-
r	Radius	L
S	Soil moisture storage	L, L^3
S_d	Soil moisture deficit	L, L^3
S_s	Soil moisture surplus	L, L^3
S_L, S_M, S_H	Soil infiltration (low, medium and high)	LT^{-1}
S_{max}	Field capacity	
S_t	Soil moisture storage at time, t	L, L^3
S_{t-1}	Soil moisture storage at time, t-1	L, L^3
T	Return period	T
T_a	Actual crop transpiration	LT^{-1}
T_c	Crop transpiration	LT^{-1}
T_L, T_M, T_H	Topography slope (low, medium and high)	-
T_m	Maximum crop transpiration	LT^{-1}
T_{CT}	Crop transpiration under traditional tillage system	LT^{-1}
T_{TT}	Crop transpiration under conservation tillage system	LT^{-1}
t	Time	T
V	Farm pond volume	L^3
Y	Crop yield	ML^{-2}
Y_a	Actual crop yield	ML^{-2}
Y_{CT}	Crop yield under conservation tillage system	ML^{-2}
Y_m	Maximum crop yield	ML^{-2}
Y_{TT}	Crop yield under traditional tillage system	ML^{-2}
η	Irrigation efficiency	-
Δ	Change (e.g. ΔY is change in crop yield)	-
θ_S	Additional soil moisture due conservation tillage (%)	-
θ_L	Non-productive additional soil moisture (%)	-

Chapter 1

1.0 Introduction

1.1 Problem Diagnosis

The semi-arid savannah environments (SASE) (commonly referred to as arid and semi-arid lands (ASAL) in Kenya) covers 83% of the country, carries 35% of the population and more than 50% of the livestock. Water is the major limiting factor to food production and socio-economic development in general. The main occupation is subsistence small-scale rainfed agriculture and livestock production, which normally compete for the limited water resources. This is aggravated by land use changes related to human settlements and intensification of agriculture in semi-arid environments, where a balance is needed between different water users and ecological functions. Moreover, there is limited knowledge on hydrological impacts on water resources management at river basin scale.

Inadequate water management due to low and poorly distributed rainfall is one of the limiting factors to improvement of water productivity in rainfed agriculture in the SASE. The problem is aggravated by poor rainfall partitioning and occurrence of intra-seasonal dry spells. According to Rockström *et al.* (2001), mitigation of intra-seasonal dry spells can be achieved through: maximizing plant water availability (maximize infiltration of rainfall, minimize unproductive water losses (evaporation, deep percolation[1] and surface runoff), increase soil water holding capacity, and maximize root depth); maximizing plant water uptake capacity (timeliness of operations, crop management, soil fertility management); and dry spell mitigation using supplemental irrigation. Analysis of field soil water balance is a useful analytical tool for assessing agricultural and water productivity in semi-arid agricultural land use systems. The challenge of increasing water productivity can be addressed by understanding rainfall partitioning and intervention points as shown in Fig. 1.1.

Rainfall partitioning analysis in Fig. 1.1 reveals that between 70-85% can be considered non-productive water flows (soil evaporation, deep percolation and surface runoff) not involved in productive plant growth. Therefore, there is a high seasonal risk of soil water scarcity in crop production, in addition to spatial and temporal rainfall variability. Reducing this risk poses one of the main challenges to upgrading rainfed agriculture in the SASE. Rockström and Falkenmark (2000) developed an analytical tool for assessing the options available to improve crop yields in semi-arid tropics from a hydrological perspective, which suggests a large scope to improve yield levels within the available water balance in rainfed agriculture. The opportunities are to maximize infiltration and soil water holding capacity, to mitigate dry spells (increase the amount of available water in relation to

---

[1] Although deep percolation is an unproductive water loss at field scale, it can contribute to groundwater recharge at a river basin scale and thus, feed other agricultural or ecosystems downstream.

crop water requirements over time) and to improve primary soil fertility management in order to increase the productive water (transpiration component). However, technologies to upgrade rainfed agriculture need to be adequately evaluated to select and promote those that improve land productivity without impacting negatively on other forms of livelihoods and natural ecosystems.

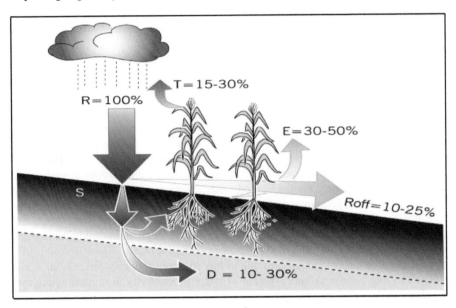

Source: Adapted from Rockström (1999)

Fig. 1.1. Overview of rainfall partitioning in agro-ecosystems in SASE. R = seasonal rainfall,
   E = soil evaporation and interception, S = soil moisture, T = plant transpiration,
   Roff = surface runoff and D = deep percolation.

Rainwater harvesting and management (RHM) is one of the viable technologies for reducing the high seasonal risk of soil water scarcity. RHM can be broadly defined as the collection, concentration/storage, management and utilization of rainwater for productive purposes. A RHM system encompasses the technology/technique/method and, biophysical and social environments. Rainwater harvesting (RWH) has been used synonymously with water harvesting. In the past, many publications referred only to RWH, however, the author instead adopts RHM. RHM puts more emphasis on the utilization and management aspects, while RWH focuses more on technical aspects of harvesting and storage.

There is an increased interest in RHM and complementary technologies for improving rainfed agriculture in SASE. Research findings (Ngigi, 2003a) indicate that RHM systems can increase crop production and reduce dependent on food relief due to frequent crop failure. However, to reduce their risk aversion, farmers in SASE need to be convinced that RHM can increase crop production and build their confidence and ultimately rescue them from the vicious cycle of poverty. Other stakeholders, especially policy makers, need to be convinced that adoption of RHM systems have no negative hydrological impacts and would not affect the livelihoods of downstream water users.

The flow chart in Fig. 1.2 shows the hydro-climatic hazards addressed by different RHM systems, and the relative socio-economic implications of implementing them in rural communities. According to Rockström (2000a), the major hydro-climatic hazards in SASE farming are: poor rainfall partitioning, where only a small fraction of rainfall reaches the root zone, coupled with within-field crop competition for soil water; high risk of periods of below optimal cumulative soil water availability during the growth season (not necessarily dry spells, but rather situations when soil water availability is below crop water requirements for optimal yields due to low cumulative rainfall levels); and high risk of intermittent droughts, or dry spells, occurring during critical crop growth stages (not necessarily a lack of cumulative soil availability, but rather periodic water stress due to poor rainfall distribution).

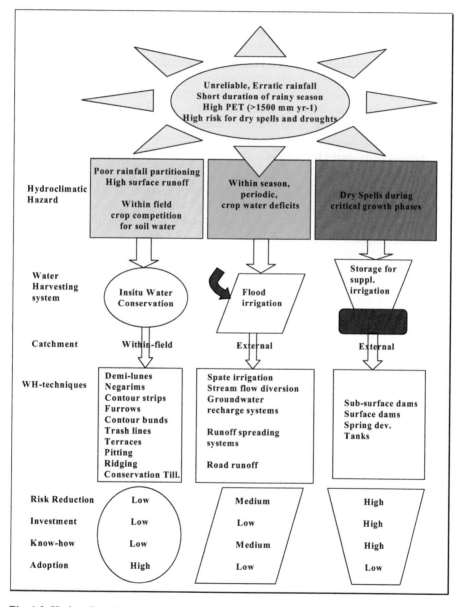

Fig. 1.2. Hydro-climatic hazards addressed by different RHM systems in SASE.

The major problem addressed by the study is spatial and temporal water scarcity for domestic, agricultural and livestock production, environmental management and overall water resources management in a river basin. Notwithstanding financial implications, solutions to this problem may have a number of inter-related options: irrigation either using water within the basin or inter-basin water transfer; RHM either in-situ soil moisture conservation, runoff farming or on-farm runoff storage for supplemental irrigation; flood diversion and storage for irrigation; soil management—improving soil fertility and water holding capacity; and crop husbandry and management, etc. However, each option cannot on its own effectively improve food security and water management in the SASE. Therefore, there is need for integrated and multi-sectoral approach to identify synergies between options in search of biophysically and socio-economically optimal systems to cope with looming water crises in water-scarce river basins. The study focused on assessment of hydro-ecological and socio-economic impacts of up-scaling RHM systems in the water-scarce upper *Ewaso Ng'iro* river basin in Kenya. The study provides an insight on the potential of RHM for improving rural livelihoods, possible impacts and what need to be considered to enhance formulation of sustainable water resources management strategies.

1.2 Hypothesis and Objectives

The hypotheses of the study were: (1) adoption of sustainable RHM systems can enhance water availability, food security and socio-economic development in water-scarce upper *Ewaso Ng'iro* river basin; and (2) up-scaling of RHM systems may have negative hydrological impacts on river flows and overall river basin water resources management.

The goal of the study was to provide information required in the formulation of integrated water resources management to enhance socio-economic development and sustain ecological balance in water-scarce river basins. The overall objective was to assess the hydrological impacts of land use changes on water resources management and socio-economic development of upper *Ewaso Ng'iro* river basin in Kenya. The specific objectives were to:

- Assess the potential of RHM technologies for improving food and water availability in semi-arid environments in Greater Horn of Africa (GHA)[2];
- Develop a conceptual framework for assessing hydrological impacts of up-scaling RHM systems in a river basin;
- Undertake agro-hydrological and economic evaluation of on-farm RHM systems in upper *Ewaso Ng'iro* river basin;
- Determine the proportion of runoff retained in agricultural catchment due to in-situ RHM systems in upper *Ewaso Ng'iro* river basin;
- Assess hydro-economic risks and options for improving and sustaining rainfed agriculture in semi-arid environments; and
- Assess the impacts of RHM on dry season's water abstractions and river flows in *Naro Moru* sub-basin of upper *Ewaso Ng'iro* river basin.

---

[2] The countries in GHA include: Burundi, Djibouti, Eritrea, Ethiopia, Kenya, Rwanda, Somalia, Sudan, Uganda and Tanzania

The study was carried out from 2002-2005, and was based on the following key issues:

- RHM would lead to improved water availability and food production—addressing the major constraints to development of SASE—low land productivity, poverty, food insecurity, land degradation and poor management of natural resources.
- RHM at farm level may reduce water availability for downstream users as more water is retained on cropland (soil profile for in-situ RHM or water pans and/or earth dams in case of on-farm storage for supplemental irrigation). However, though RHM may seem to reduce water flow downstream, this may not have a major impact since water is "abstracted" during the rainy seasons—flood diversion and storage during high flows.
- RHM can improve soil management at field and watershed scale through minimizing soil erosion as runoff flow velocity and volume are reduced as its proportion is retained in the watershed. This will reduce sediment loads, eutrophication of wetlands ecosystem and hence sustain soil fertility and land productivity.
- RHM is being promoted and adopted in the upper *Ewaso Ng'iro* river basin, through farmers' initiative and development partners' supported programs. However, there is a lot that need to be done to improve these systems and promote adoptability and adaptability to realize their potential for enhancing rural development and livelihoods.
- RHM can supplement irrigation and may reduce irrigation water requirements and hence increases water availability for downstream users. The water saved at the farm level may translate into real savings at the river basin level.
- The need to develop sustainable strategies to address looming water crisis and conflicts among different stakeholders, in the view of increasing water abstractions (RHM and irrigation) upstream and decreasing river flows for downstream users and ecological use.

1.3 Thesis Outline

The Ph.D. dissertation is presented in eight chapters. Chapter 1 is the introduction and provides an overview of problem diagnosis, hypotheses and objectives, keys issues that form the basis of the study, and thesis outline. Chapters 2-7 are based on six journal papers, of which four are published while the remaining two have been submitted for publication. Chapter 2 outlines the research background, description of the study area and an overview of research methodology. It focuses on results of review and analysis of different RHM systems in GHA. Chapter 3 presents a conceptual and analytical framework for assessing hydrological impacts of land use changes, in particular adoption of RHM systems and irrigation at a river basin scale, and forms the basis of the other chapters.

Chapter 4 presents agro-hydrological assessment of on-farm storage RHM system, which focuses on evaluation of small on-farm storage systems (30-50m³ farm ponds) for growing vegetables under supplementary irrigation on a kitchen garden scale. An economic analysis of the system was also carried out. Chapter 5 presents agro-hydrological assessment in-situ RHM (conservation tillage) systems, which focuses on assessment of the proportion of rainwater retained on agricultural land due to adoption of conservation tillage, which enhances additional soil moisture storage for plant use. An analytical approach, based on increased grain yields, for estimating the additional amount of rainwater retained in the soil profile and used by the plant is developed and compared with field measurements.

Chapter 6 presents an analysis of hydro-economic assessment and farmers' investment options. It highlights hydrological hazards and risks affecting farmers in SASE and some of the viable options for improving and upgrading rainfed agriculture under water scarcity conditions. It outlines the role played by RHM, especially on-farm runoff storage ponds, in reducing the impacts of intra-seasonal and off-season dry spells through supplemental irrigation. Results of economic analysis for maize under traditional and RHM systems are presented. It also explores investment recovery time in case a farmer borrows money to investment in on-farm storage systems for supplemental irrigation.

Chapter 7 presents an assessment of hydrological impacts of flood storage on dry season's water abstraction at a river sub-basin scale and its implication on river flows at the basin scale. It is based on the analysis of long term river flow records and periodical monitoring of water abstractions in *Naro Moru* sub-basin. The hydrological impacts of flood storage was assessed using the naturalized river flows for a below average year (i.e. 2002) when the demand for irrigation water is highest. The impact of different levels of flood storage is assessed in terms of reduced high flows and low flow irrigation water abstraction. Lastly, chapter 8 consolidates conclusions from the six papers and presents some recommendations towards formulation of a sustainable water resources management strategy in water-scarce river basins.

Chapter 2

2.0 Research Background and Study Area

2.1 Research Background [3]

2.1.1 Overview

The SASE in sub-Saharan Africa (SSA) are characterized by low erratic rainfall which result to high risk of droughts, intra-seasonal dry spells and frequent food insecurity. Water is one of the limiting factors to food production and socio-economic development in general. The main occupation is subsistence small-scale rainfed agriculture and livestock production, which normally compete for the limited water resources. The main challenges to improving the livelihoods of the small-scale farmers are how to upgrade rainfed agriculture to improve rural livelihoods and conserve nature, and upgrade upstream land use in balance with water needs for human and ecosystems downstream. There is an increased interest in opportunities of improving rainfed agriculture through adoption of RHM systems. However, there is inadequate knowledge on hydrological impacts and limits of up-scaling RHM at a river basin scale.

RHM systems can address spatial and temporal water scarcity for domestic, crop production, livestock development, environmental management and overall water resources management in semi-arid areas. However, this potential has not been exploited despite the occurrence of persistent low agricultural production and food shortage in SSA. The need to quantify the perceived potential and related hydrological impacts on a river basin formed the basis of the study. It is envisaged that the study will contribute to formulation of sustainable RHM up-scaling strategies to enhance food production and hydro-ecological balance in SASE of Africa.

The section presents the preliminary findings mainly focusing on assessment of the potential of RHM technologies for improving food and water availability especially in semi-arid regions of eastern Africa. This was achieved by evaluating six RHM case studies selected from four countries (Ethiopia, Kenya, Tanzania and Uganda). The case studies were based on participatory evaluation in which the land users directly participated in reviewing their systems, identifying shortcomings, proposing possible solutions, analyzing various alternatives of addressing the shortcomings, identifying viable and feasible solutions, and adapting and adopting promising RHM systems. The case studies evaluated some of the constraints and opportunities that the land users experience in their endeavours to address persistent food insecurity and water scarcity.

---

[3] *Based on*: Ngigi, S.N. 2003. What is the limit of up-scaling rainwater harvesting in a river basin? *Physics and the Chemistry of the Earth, 28: 943-956*

The case studies revealed that many solutions related to adoption of rainwater technologies can be developed by the land users. There are various promising aspects of RHM technologies, geared towards improving subsistence food production, being adopted by rural land users. Moreover, there are disparities in the rate of adoption and adaptation, and types of technologies in different countries and even regions within a country. Despite the success of a number of RHM systems, the rate of adoption is still low, hence making their impacts marginal. For instance, it is common to find a farmer producing substantial yield by adopting RHM technology while the neighbour's crop has completely failed during droughts. Therefore, promotion and adoption of promising RHM practices could address the recurrent food crises in SSA.

Nevertheless, there is a knowledge gap on the limits of up-scaling RHM in a river basin scale. The assessment of the hydrological impact of up-scaling RHM systems can be done using hydrological model(s) that simulate the effects of different adoption rates of RHM technologies in a river basin, i.e. hydrological impacts downstream/upstream. The effect of RHM on upstream and downstream water resources management and impacts on the ecosystem was assessed. It is envisaged that the results will provide insight on the potential of RHM for improving food security, and the sustainable level (threshold) of up-scaling the technologies without affecting river basin water resources management. Thus it will provided answers to the question, *what is the limit of up-scaling rainwater harvesting in a river basin?*

Most of the countries in SSA are experiencing profound socio-economic and political problems, the most dramatic being food crises and disruptive conflicts. The communities involved are experiencing a combination of both short-term, often acute food crises, and long-term or chronic food shortages. The former often translate into famine and starvation, requiring emergency food aid. The latter are less obvious, for they are characterized by negative changes in the economic, social and ecological factors and their interrelationships over longer time periods. These crises threaten the stability and existence of the affected communities and economies because their systems are obviously failing to cope, increasing the vulnerability of the people. A number of explanations have been advanced for the endemic food insecurity in the SSA. Among these, recurring drought and unreliable rainfall are the most obvious. These include: adverse weather and drought; rapid population growth rates that exceed rates of food production; adoption of production systems that accelerate environmental degradation and decline in soil fertility; and retrogressive social organizations, inadequate policies, legislation and institutional weaknesses.

Over 60% of the land in the SSA falls under SASE, where a majority of the inhabitants are pastoralists although agro-pastoral and farming communities have been slowly settling in these areas due to population pressure in the high agricultural potential areas. SASE is predominantly characterized by low and variable rainfall, which rarely exceeds 800 mm yr^{-1}, with most areas receiving 200-350 mm yr^{-1}. The water resources are limited and poorly distributed. There are few permanent rivers, and seasonal streams that flow only during the wet season and remain dry for the rest of the year. However, like the wetter regions, SASE too is starting to experience land pressure resulting from population increase within the communities and also their livestock. This has significantly raised pressure on pastures leading to overgrazing and decline in vegetation cover in most of the areas. The impact of the frequent droughts that hit the pastoral areas has therefore been increasing over time. Huge livestock numbers have been dying every time there is drought (Kihara 2002). Much of the pastureland has lost grass cover and is often bare. This leaves the people highly vulnerable. Consequently the SASE form specific pockets of poverty and food insecurity, and ensuing conflicts, especially over diminishing natural resources—mainly water and pasture.

Nevertheless, unreliable rainfall and low soil fertility has continued to threaten food production in the SSA thus making food security a major concern. Currently, vast areas of SSA are facing drought and the threat of famine despite the fact that overall food production could be adequate. Relief food has on many occasions saved lives in the region from severe famine situations. Food relief will continue to be required as long as transportation facilities are poor and local food production in drought prone areas is inadequate. Given the poor transportation infrastructure, emphasis on local food production appears the most logical approach to improved food security.

Agriculture is the major economic activity for the countries of the SSA, engaging between 75% and 85% of the people of those countries. Consequently, it is strongly underscored that agriculture is the backbone of these countries' economic development and their people's well being in the foreseeable future. A survey of 277 societies in SSA by Hunt-Davis (1986) showed that approximately 86% depended primarily on agriculture, 6% on animal husbandly, and animal husbandry and agriculture are co-dominant for another 3%. Of the rest, 2% rely primarily on fishing; 1% on fishing and agriculture equally, and some minorities on hunting and gathering. Thus the livelihood in this region is based on small-holder rural agriculture, with low levels of productivity and simple tools, making them over-dependent on the status of the natural environment. Seasonal rainfall dominates the lives of most of the people, as it determines their activities geared towards earning a livelihood based on exploitation of the resources of the land. Duckham and Masefield (1985) stated that in the tropics generally, rainfall is the main determinant of agricultural activities. The same fact had been expressed by Jodha and Mascarenhas (1985) as characteristic of much of the rest of Africa. In GHA, rainfall—amount, timing, duration and distribution—was identified by subsistence farmers as the main determining factor for food production and security.

Therefore, the problems related to food security and recurrent famine need urgent solutions, especially in the SASE, where environmental degradation has further decreased agricultural productivity, making inhabitants even more susceptible to drought and other natural disasters. Unless sustainable food production technologies are adopted, alleviation of poverty and food security will remain elusive. RHM is one of the promising technologies for improving food production. This is the process of interception and concentration of runoff and its subsequent storage in the soil profile or in artificial reservoirs for crop production. The process is distinguished from irrigation by three key features: the catchment area is contiguous with the cropped area and is relatively small; the application to the cropped area or reservoir is essentially uncontrolled; and water harvesting can be used for purposes other than crop production. There are many techniques being used to enhance crop production in the SASE of the SSA. However, the viability of these solutions needs to be evaluated in relation to environmentally sustainable factors, climatic conditions, soil characteristics, farming systems and socio-cultural and gender perspectives in which they are practiced.

Participatory evaluation is needed to determine viable options and adaptive strategies for sustainable food production in the SASE. Needless to say, the solutions must be land user-oriented, hence the need for a participatory technology development approach. The project gave special attention to RHM systems, which are being used by land users. Any activity to improve on land users' innovations and the applicability of those innovations will be a major contribution to food security in this famine prone region. It is with a sense of urgency that one notes the relevance of RHM technologies in certain limited but significant areas of Africa, both for food production and for soil and water conservation (Pacey and Cullis, 1986). Despite this realization, little practical information exists on RHM technologies, which can be applied on site specific situations. RHM is one of the approaches to integrated land and water management, which could contribute to recovery of agricultural production in dry areas.

2.1.2 Food Production and Water Scarcity in SSA

The semi-arid areas of SSA are characterized by low annual rainfall concentrated to one or two short rainy seasons. The average annual rainfall varies from 400-600 mm in the semi-arid zone, and ranges between 200-1,000 mm from the dry semi-arid to the dry sub-humid zone (Rockström, 2000a). The length of the growing period ranges from 75-120 days and 121-179 days in the semi-arid zone and dry sub-humid zone respectively. Potential evaporation levels are high, ranging from 5-8mm/day (FAO, 1986) giving a cumulative evapotranspiration of 600-900mm over the growing period. This explains the persistence water scarcity coupled with low crop yields. Water scarcity could also be attributed to poor rainfall partitioning leading to large proportion of non-productive water flows—not available for crop production. The nature and occurrence of rainfall in SASE of SSA provides more insight in the food production and water scarcity situations.

Rainfall is highly erratic, and normally falls as intensive storms, with very high intensity and spatial and temporal variability. The result is a very high risk for annual droughts and intra-seasonal dry spells (Rockström, 2000a). From past experience, severe crop reductions caused by dry spells occurs 1-2 out of 5 years, while total crop failure caused by annual droughts occur once in every 10 years in semi-arid SSA. This means that the poor distribution of rainfall, more often than not, leads to crop failure than absolute water scarcity due to low cumulative annual rainfall. Unfortunately, most dry spells occur during critical crop growth stages (this explains frequent crop failure and/or low yields), and hence the need of dry spell mitigation by improving water productivity in SSA.

From the above brief overview of rainfall patterns, there is a growing understanding that the major cropping systems in SSA are not sustainable (Benites *et al.*, 1998), hence the persistence low food production (food shortage) and reliance on food relief. The livelihoods of majority of the population in SSA is based on rainfed agriculture, and depend on to a large extent on smallholder, subsistence agriculture for their livelihood security (e.g., Botswana, 76%; Kenya, 85%; Malawi, 90%; and Zimbabwe, 70-80% of the population (Rockström, 1999). Moreover, an estimated 38% of the population in SSA roughly 260 million people lives in drought prone SASE (UNDP/UNSO, 1997). This may explain why most of the population is poor—rely on unsustainable farming systems, majority living below the poverty limit (< US$ 1 per day). The key role of agriculture in Africa's economic life is apparent—agriculture accounts for 35% of the continent's GDP, 40% of its export, 70% of its employment, and more than 70% of the population depend for their livelihoods on agriculture and agri-business (Kijne, 2000).

The problem of low food production is further aggravated by limited new land for cultivation, land and environmental degradation, poor infrastructure, political and social crises, bad governance, insecure land tenure, health/diseases outbreak, inadequate knowledge/capacity, and donor dependency syndrome. Thus the ever increasing food demand and household income needed in SSA have to be achieved through an increase in biomass produced per unit land and unit water (Rockström, 1999). In the past, very little attention has been paid to the development of rainfed agriculture in SSA except provision of conventional irrigation projects. However, most of these projects have proven (e.g. *Bura* irrigation scheme in Kenya) to be unnecessary, costly and environmentally unsustainable. Hence the need to focus on opportunities of increasing efficiency of limited water in rainfed, smallholder agriculture in the SASE of SSA. Otherwise feeding the ever growing population (at a rate of 2-3% per year) with diminishing crop yields (oscillating around 1 ton/ha for food grains) in SSA (Rockström, 2001) will remain elusive, and current generation's biggest challenge.

2.1.3 RHM Systems

RHM is broadly defined as the collection, concentration and management of runoff for productive purposes (crop, fodder, pasture or trees production, livestock and domestic water supply). It has ancient roots and still forms an integral part of many agricultural systems worldwide (Evanari *et al.*, 1971; Shanan and Tadmor, 1976; Critchley, 1987; Critchley and Siegert, 1991; Agarwal and Narain, 1997). It includes all methods of concentrating, diverting, collecting, storing, and utilizing and managing runoff for productive use. However, in-situ systems i.e. on-farm/cropland water conservation—to enhance soil infiltration and water holding capacity—dominate, while storage systems for supplemental irrigation are less common, especially in SSA (SIWI, 2001). Nevertheless, a recently concluded evaluation of RHM in four GHA countries (i.e. Ethiopia, Kenya, Tanzania and Uganda) revealed that, despite the relatively high investment costs compared to in-situ systems, RHM for supplemental irrigation is slowly being adopted with high degree of success (Kihara, 2002). In this system, surface runoff from small catchments (1-2 ha) or adjacent road runoff is collected and stored in manually and/or mechanically dug farm ponds (50-1,000m^3 storage capacity). Due to the low volumes of water stored compared with crop water requirements, improved benefits of these systems are derived by incorporating efficient water application methods such as low pressure (0.5-1.5m) drip irrigation (Ngigi *et al.*, 2000; Ngigi, 2001; Ngigi, 2002b).

Furthermore, on-farm research in semi-arid locations in Kenya (Machakos district) and Burkina Faso (*Ouagouya*) during 1998-2000 (Barron *et al.*, 1999; Fox and Rockström, 1999) indicates a significant scope to improve water productivity in rainfed agriculture through supplemental irrigation, especially if combined with soil fertility management. The results were more promising on soils with higher water holding capacity on which crops seem to cope better with intra-seasonal dry spells. However, incremental water productivity improvements are only achieved during rainy seasons with severe dry spells, while rainy seasons with adequately distributed rainfall the incremental value can be negative (Rockström *et al.*, 2001).

Runoff is collected mainly from ground catchments as well as ephemeral streams (flood water harvesting) and road/footpath drainage. The storage is either in different structures (tanks, reservoirs, dams, water pans, etc.), mainly for supplemental irrigation systems, or soil profile (for in-situ and flood irrigation). RHM can be considered as a rudimentary form of irrigation (Fentaw *et al.*, 2002). The difference is that with RHM the farmer has no control over timing, as runoff can only be harvested when it rains. In regions where crops are entirely rainfed, a reduction of 50% in the seasonal rainfall, for example, may result in a total crop failure (Critchley and Siegert, 1991). However, if the available rain can be concentrated on a smaller area, reasonable yields will still be received. Fig. 2.1 shows the principle of RHM, which is common for different classifications, except in-situ (no runoff) systems which capture rainfall where it falls. Classification of runoff-based RHM technologies depends on:

- Source of runoff (external) or within-field catchments;
- Methods of managing the water (maximizing infiltration in the soil, storing water in reservoirs and inundating cropland with floods); and
- Use of water (domestic, livestock, crop production, gully rehabilitation, etc.).

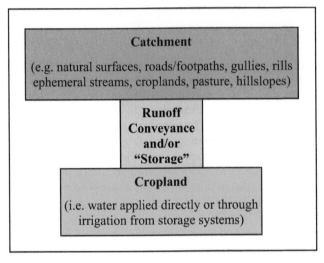

Fig. 2.1. The principle of a runoff-based RHM system

RHM systems operate at different scales (household, field and catchment/basin), and can affect water availability and management for downstream and natural ecosystems like wetlands and swamps, due to reduced catchment water yields. Therefore, even though RHM practices can be efficient in increasing the soil moisture for crops (principle objective) in water scarce areas, each technique has a limited scope due to hydrological and socio-economic limitations. Rockström (2000a) highlighted the major hydro-climatic hazards in SASE farming as:

- Poor rainfall partitioning, where only a small fraction of rainfall reaches the root zone, coupled with within-field crop competition for soil water;
- The high risk of periods of below optimal cumulative soil water availability during the growth season (i.e. not necessarily dry spells, but rather situations when soil water availability is below crop water requirements for optimal yields due to low cumulative rainfall levels); and
- The high risk of intermittent droughts, or dry spells, occurring during critical crop growth stages (i.e. not necessarily a lack of cumulative soil availability, but rather periodic water stress due to poor rainfall distribution.

2.1.4 RHM Systems in GHA

This section briefly presents the different RHM systems found in SSA, especially in GHA, focusing on their classifications, and their opportunities and limitations for improving rainfed agriculture in SASE. The term RHM is used in different ways and thus no universal classification has been adopted. However, according to Oweis *et al.* (1999) the following are among its characteristics: RHM is practiced in SASE where surface runoff is intermittent; and is based on the utilization of runoff and requires a runoff producing area (catchment) and a runoff receiving area (cropped area and/or storage structures). Therefore, each RHM system, except in-situ water conservation (see Fig. 2.1) should have the following components: runoff producing catchment, runoff collection (diversion and control) structures, and runoff storage facility (soil profile in cropland or distinct structure (farm ponds, tanks, water pans, earthdams, sand dams, subsurface dams, etc.).

To avoid further confusion, and facilitate the presentation of various types of RHM systems, the classification shown in Fig. 2.2 is based on runoff generation process, type of storage/use and size of catchments was adopted. The runoff generation criteria yields two categories—runoff farming (where runoff is generated i.e. runoff-based systems) and in-situ water conservation (rainfall conserved where it falls). The runoff storage criteria also yields two categories— soil profile storage (direct runoff application) and distinct storage structures for supplemental irrigation, livestock, domestic or commercial use). Whilst the size of catchment criteria yields three categories—macro-catchments (flood diversion and spreading i.e. spate irrigation), small external catchments (road runoff, adjacent fields, etc.), and micro (within field)-catchments (e.g. *Negarims*, pitting, small bunds, tied ridges, etc.)[4].

Moreover, runoff storage structures capture runoff mainly from small catchments especially for small-scale land users, but macro-catchments with large storage structures could also be used for large-scale or community-based projects. In-situ water conservation could also be considered under soil profile storage systems, only that in that case direct rainfall is stored, but not surface runoff. However, the classification is further complicated by the fact that a number of RHM technologies are integrated or combined by land users, for example, fields under conservation tillage in *Laikipia* district also incorporate runoff spreading from small external catchments such as road/footpath drainage and adjacent fields.

It is also common to find runoff from external catchment being directed into cropland with farm ponds for supplemental irrigation. In-situ water conservation is also combined with runoff farming on farms with terraces, in which the terrace channel (mainly *fanya juu* and contour ridges/bunds) collects and stores runoff from small external catchments while the cropland between the channels harvest and conserve direct rainfall. However, excess runoff that may be generated from the cropland between the terrace channels would be collected at the channel.

The following sub-sections highlight some of the RHM systems that have been tried, experimented and practiced in different parts of GHA, in addition to those identified and evaluated as part of the case studies in parts of Ethiopia, Kenya, Uganda and Tanzania.

---

[4] Micro-catchment also refers to field scale catchment (1-5ha), while small catchment refers to as hillslope or medium scale catchment (100-200ha). Macro-catchment refers to large catchment, i.e. sub-basin or river basin scale (>20 km^2).

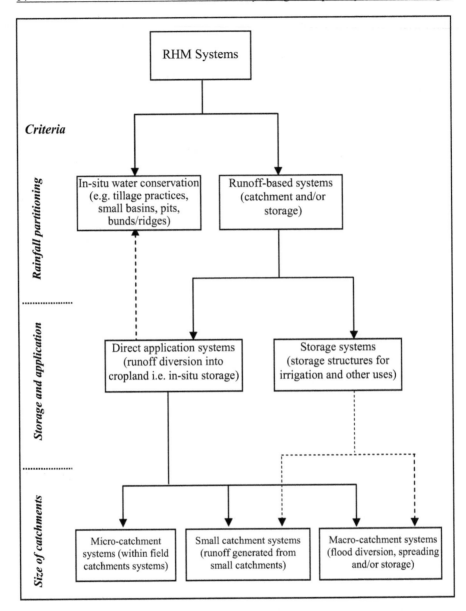

Fig. 2.2. Adopted classification of RHM systems in GHA

In-situ rainwater conservation

In-situ rainwater conservation technologies are distinct from runoff farming systems in that they do not include a runoff generation area, but instead aims at conserving the rainfall where it falls in the cropped area or pasture. The most common technology is conservation tillage which aims to maximize the amount of soil moisture within the root zone. A number of cultural moisture practices such as mulching, ridging, addition of manure, etc. could fall under this category. Small field/farm structures such as tied ridges/bunds within cropped area that conserve direct rainfall without 'external'—outside cropland boundary, i.e. no distinct catchment area, except overflow from upstream sections also falls under this category. Within cropland or pasture contour bunds/ridges, bench terraces, and sweet potatoes ridges practiced in *Rakai* district of Uganda could also fall under this category.

In-situ rainwater conservation technology is one of the simplest and cheapest and can be practiced in almost all the land use systems. In-situ water conservation systems are by far the most common (Rockström, 2000a) and are based on indigenous/traditional systems (Reij *et al.*, 1996; LEISA, 1998). The primary objective has been to control soil erosion and hence manage the negative side-effects of runoff—soil and water conservation measures, i.e. ensures minimal runoff is generated. The positive effect of in-situ water conservation techniques is to concentrate within-field rainfall to the cropped area. In a semi-arid context, especially with coarse-textured soil (especially sandy soils common in the SASE) with high hydraulic conductivity, this means that in-situ conservation may offer little or no protection against the poor rainfall distribution. In such cases, the farmers will continue to live at the mercy of the rain. In effect, the risk of crop failure is only slightly lower than that without any measures. However, soil improvements and management would enhance realization of better yields.

Conservation tillage

Conservation tillage is defined as any tillage sequence having the objective to minimize the loss of soil and water, and having an operational threshold of leaving at least 30% mulch or crop residue cover on the surface throughout the year (Rockström, 2000a). However, with respect to small-scale farmers in SASE, conservation tillage is defined as any tillage system that conserves water and soil while saving labour and traction needs. Conservation tillage aims at reversing a persistent trend in farming systems of reduced infiltration due to compaction and crust formation and reduced water holding capacity due to oxidation of organic materials (due to excessive turning of the soil). From this perspective, conservation tillage qualify as a form of water harvesting, where runoff is impended and soil water is stored in the crop root zone (Rockström *et al.*, 1999).

Unlike the conventional tillage systems, based on soil inversion which impedes soil infiltration and root penetration, conservation tillage covers a spectrum of non-inversion practices from zero-tillage to reduced tillage which aim to maximize soil infiltration and productivity, by minimizing water losses (evaporation and surface runoff) while conserving energy and labour. Kihara (2002) revealed the successes of conservation tillage in harnessing rainwater and improving yields. Field visits in *Machakos* revealed that, during the recent below average short rains (2001/2002), farms where conservation tillage was practiced had good harvest while adjacent farms without convention tillage had literally no harvest—conspicuous contrast.

Conservation tillage has several attractive effects on water productivity (Rockström *et al.*, 2001) compared to traditional soil and water conservation systems such as *fanya juu* terracing in *Machakos* district. In addition to enhancing infiltration and moisture conservation, it enables improved timing of tillage operations, which is crucial in semi-arid rainfed farming. It can also be applicable on most farmlands compared say to storage RHM systems for supplemental irrigation. Promotion of animal drawn conservation tillage tools such as rippers, ridgers and sub-soilers among smallholder farmers in semi-arid *Machakos* and *Laikipia* districts (Kenya) has resulted in significant water productivity and crop yields (Kihara, 2002; Muni, 2002). There are many documented examples of successful conservation tillage practices in ESA, where crop yields have been increased through the conservation of soil water and nutrients and/or draught power needs have been reduced (Rockström *et al.*, 1999).

The findings of the case studies in *Laikipia* and *Machakos* districts of Kenya reveals that conservation tillage (sub-soiling and ridging) have improved yields by more than 50% (Kihara, 2002; Muni, 2002). The potential of conservation tillage is tremendous especially with communities already using animal drawn implements for their tillage operations. This is because conservation tillage implements are compatible with the conventional tools. On large scale farming systems in Laikipia district, tractor drawn conservation tillage implements have improved wheat yields. Pastoral communities are also not being left behind, as ground scratching using animal or tractor drawn tools have improved pasture development in *Laikipia*.

In *Dodoma*, Tanzania, trench cultivation, a form of conservation tillage have been developed by innovative farmers, where shallow trenches are dug, filled with organic materials then covered by soil to form ridges on which crops are planted (Lameck, 2002). The 'organic' furrows between the ridges capture water, which seeps into the covered trenches and is slowly extracted by the crops. The organic material improves soil fertility and water holding capacity. This seems to be an improvement of the furrow and ridge systems as used in *Kitui* and *Machakos*. The furrows and ridges are made using animal drawn mould board ploughs. Seeds are planted in the furrow, which collects water between the ridges. After seedlings develop, weeding operation (using animal drawn ridgers) ensures that the furrows and ridges alternate—the crops grow on the ridges while the furrows captures and concentrates the rainwater. In trench cultivation, the ridges and furrows are rotated after each season and have enhanced crop yields in otherwise low yielding areas.

Runoff-based RHM systems

The runoff-based RHM systems, which entail runoff generation either within field or from external catchments and subsequent application either directly into the soil profile or through periodic storage for supplemental irrigation, are classified according to two criteria (as shown in *Fig. 2*): runoff storage and/or application, and size of catchments.

Storage RHM systems

RHM systems with storage for supplemental irrigation are becoming popular in semi-arid districts of Kenya (e.g. *Machakos*, *Laikipia*, and *Kitui*). They have also been introduced in central Ethiopia (near *Nazareth*) on experimental basis by RELMA. Moreover, small storage systems are all over parts of Ethiopia (e.g. *Tigray*), and other places around Africa (Merrey, 2002). Initial results from RHM experiments in *Machakos* district, which focused on the feasibility of using earthdams for supplemental irrigation of maize have been encouraging (Rockström *et al.*, 2001). However, the main challenge is to assess whether it is possible to design simple and cheap earthdams or farm ponds that could permit gravity-fed irrigation to reduce the cost of lifting water.

In the semi-arid parts of *Laikipia* district (Kenya), underground water tanks ($50\text{-}100\text{m}^3$ capacity) have been promoted mainly for kitchen gardening. The tank surfaces are usually sealed with polythene lining, mortar, rubble stones or clay to reduce seepage losses while covering the tanks, with either local material (thatch or iron sheet), minimizes evaporation. However, similar initiative in *Kitui* district was discouraging as most of the mortar sealed underground tanks ended up cracking and hence being abandoned (Ngure, 2002). In *Laikipia*, loss of water through seepage has been identified as a major drawback (Kihara, 2002). Thus despite the positive impact realized by this technology, its widespread adoption could be hampered if simple seepage control measures are not devised. Concrete sealing seem to work well in *Ng'arua* division of *Laikipia* district, but the cost may be beyond the reach of many farmers. Farmers are still experimenting with various seepage control methods, among them, plastic lining (found not durable), butimen lining, clay lining and even goats trampling. Nevertheless, seepage control still remains a major challenge.

Other storage systems used by small-scale farmers in semi-arid districts of eastern Kenya are rock catchments/dams, sand dams and sub-surface dams (Gould and Petersen, 1999; Pacey and Cullis, 1986). Sand dams and subsurface dams are barriers constructed along sandy riverbeds—a common phenomenon in most semi-arid environments in GHA—to retain water within the trapped sand upstream. These systems have provided water for decades especially in *Machakos* and also in some parts of *Kitui* district. They have also been introduced in the *Dodoma,* Tanzania but their potential has however, not been realized. They provide water for all purposes and could lead to environmental improvement, for example in *Utooni* in *Machakos* and some parts of central *Kitui*. The impacts of sand dams on food security have been highlighted by Isika *et al.* (2002). They are mainly used for domestic purposes, but in several cases also used for small-scale irrigation (Rockström, 2000a). Rock catchment dams are masonry dams, for capturing runoff from rock surfaces/catchments, with storage capacities ranging from 20-4,000m^3. They are generally used for domestic purposes, but can also be used for kitchen gardening, for example in *Kitui* district (Ngure, 2002).

RHM storage systems offers the land user a tool for water stress control—dry spell mitigation. They reduce risks of crop failures, but their level of investment is high and requires some know-how especially on water management. However, these systems also to some extent depend on rainfall distribution. During extreme drought years, very little can be done to bridge a dry spell occurring during the vegetative crop growth stage if no runoff producing rainfall have fallen during early growth stages. Under normal intra-seasonal droughts, the farmer will be assured of a better harvest and hence it is worthy any investment to improve crop production in the semi-arid tropics of SSA. Nevertheless, location of the storage structure with respect to cropland needs to be addressed. Conventionally, the reservoirs are located downstream, thus requiring extra energy to deliver the water to the crops. However, it would be more prudent to locate the reservoir upstream of the cropland to take advantage of gravity to deliver the water (Rockström, 2000a).

Runoff is collected from grazing land, uncultivated land, cultivated land and road drainage and directed into small manually constructed reservoirs (50–200m^3). The stored water is mainly utilized for kitchen gardening and establishment of orchards. This technology was introduced in *Laikipia* district Kenya in the late 1980's by the Anglican Church of Kenya and has shown promising results. It has been promoted by Dutch-supported ASAL and SARDEP progammes with limited success due to seepage related problems. Various remedies are being tried to reduce seepage to realize maximum benefits from this technology. In Kenya, it has also been introduced in *Machakos* district by RELMA. Optimal benefits could be realized if appropriate water lifting and application technologies such as treadle pump and drip irrigation are incorporated. Farm ponds have also been used for watering livestock. At community level, earthdams or water pans are constructed to store large quantities of water, especially for livestock and small-scale irrigation. These water pans and earthdams are the lifeline for livestock in the ASAL of Kenya, Somalia and parts of Uganda (southern and northeastern). The earthdams were introduced by white settlers while the water pans have been traditional sources of water e.g. *hafir*s (water pans) in North Eastern Province of Kenya, parts of Somalia and western Sudan (Critchley, 1987). Concrete/mortar lined underground tanks (100-300m^3) are used for domestic and some livestock (milking cows, calves or weak animals, separated from the main herds) in Somaliland (Pwani, 2002).

Direct runoff application RHM systems

This category of RHM technology is characterized by runoff generation, diversion and spreading within the cropland, where the soil profile acts as the moisture storage reservoir. This technology is further classified, according to size of catchments: macro-catchments systems—large external catchments producing massive runoff (floods) which is diverted from gullies and ephemeral streams and spread into cropland, i.e. spate irrigation; small external catchments (e.g. road drainage, adjacent fields, etc.) from which runoff is diverted into cropland; and micro-catchments normally within cropland which generate small quantities of runoff for single crops, group of crops or row crops.

Flood diversion and spreading (spate irrigation) systems

Flood diversion and spreading (i.e. spate irrigation) refers to RHM system where surface runoff from macro-catchments concentrating on gullies and ephemeral streams/water courses is diverted into cropped area and distributed through a network of canals/ditches or wild flooding and subsequently retained in the field by bunds/ridges. It entails controlled diversion of flash floods from denuded highlands to cropped land well prepared to distribute and conserve the moisture within the plants rootzone. The rainfall characteristics in the semiarid savannah environment occurs as high intensity storms that generate massive runoff that disppear within a short period through seasonal waterways. Worse still the number of rainstorms are normally limited within the short rainy seasons.

Extensive flood irrigation of paddy rice in cultivation basins (commonly referred to as "*majaluba*") created from 25-100cm high earth bunds, is practiced in semi-arid central parts of Tanzania (*Dodoma, Singida* and *Shinyanga*) (Mwakalia and Hatibu, 1992; Hatibu *et al.*, 2000; Lameck, 2002). It is estimated that 32% of Tanzania's rice production originate from cropland where RHM is practiced. Similar techniques have been used for maize and sorghum in Tanzania.

Spate irrigation in northern Ethiopia and Eritrea, involve capturing of storm floods from the hilly terrain and diversion into leveled basins in the arid lowlands croplands. In *Kobo Wereda* (south of *Tigray*), spate irrigation is well developed with main diversion canals, secondary/branch, tertiary and farm ditches which distribute flood water into cultivation basins with contour bunds to enhance uniform water application. A series of main canals for different group of farmers are normally constructed together to reduce destruction by floods (Fentaw *et al.*, 2002). Farmers in *Kobo* plains in northern Ethiopia have developed a traditional irrigation system that diverts part of such floods to their farms. These system have sustained livelihoods that would otherwise be impossible in that dry part of the country. These systems are similar in principle to those developed by the early settlers of the *Negev* desert in Israel. The system has also been tried in *Konso*, southern Ethiopia. This technology has also been practiced in *Turkana* district, Kenya for sorghum production and parts of Sudan (Cullis and Pacey, 1992). In western Sudan, terraces and dykes are used for spreading runoff from *wadis/laggas* onto *vertisols* (Critchley, 1987).The potential of these systems are enormous and if improved and promoted could lead to food security.

The use of external catchments for runoff collection immediately adds water to the field scale water balance. With flood irrigation systems in the SASE where absolute crop water scarcity is common, crop yields can be improved substantially during years with reasonably good rainfall distribution. The farmers still live under the mercy of the rains, but when it rains, the supply of water to the root zone exceeds rainfall depths. This can be addressed by introducing storage facilities.

Small external catchment systems

These include a form of small-scale flood/runoff diversion and spreading either directly into cropland or pasture through a series of contour bunds or into terrace channels and other forms of water retention structures. The runoff is either conveyed through natural waterways, road drainage or diversion/cutoff drains. Road/footpath runoff harvesting is practiced in parts of Kenya (*Machakos* and *Laikipia*), in which flood water from road/footpath drainage is diverted either into storage for supplemental irrigation or into croplands (wild flooding, contour bunds, deep trenches with check-dams to improve crop yields. Similar system is practiced in southwestern Uganda, where runoff from gullies, grazing land, or road drainage is diverted into banana plantations (Kiggundu, 2002).

Fanya juu terraces which were previously used with diversion/cutoff drains for soil conservation, especially in *Machakos* and *Kitui,* have been adopted as in-situ RHM system. They are modified by constructing planting pits mainly for bananas and tied ridges (check dams) for controlling the runoff. The outlet is blocked to ensure as much runoff as possible is retained while spillways are provided to discharge excess runoff, which is normally diverted into the lower terraces. Runoff spreading has also been accomplished by contour bunds in *Laikipia* district. They collect and store runoff from various catchments including footpaths and road drainage. The stored runoff seeps slowly into lower terraces ensuring adequate moisture for crops grown between the terrace channels. In southern Uganda, a similar system has been adopted, in which contour ridges/bunds, (shallow *fanya juu* terraces) tied at regular intervals are used in banana plantations. The runoff from hilly grazing lands is distributed into the banana plantations by contour ridges.

Agroforestry (for firewood and fodder) is also incorporated, where trees are planted on the lower side and Napier or giant Tanzania grass along the ridges. This system has tremendously improved the yield of the bananas and has enhanced zero grazing. Contour ridges and infiltration trenches have also been adopted to improve pasture in southern Uganda (Kiggundu, 2002). The infiltration trenches are dug at specified intervals according to the land slope and tied at regularly intervals to allow water retention and subsequent infiltration. The soil is either thrown upward (*fanya juu)* or downwards (*fanya chini*) and stabilizing grass or fodder crops. Runoff from uphill catchments is normally diverted into these contour ditches (infiltration trenches) to increase soil moisture.

In eastern Sudan, a traditional system of harvesting rainwater in "terraces" is widely practiced (Critchley, 1987). It consists of earthen bunds with wing walls which impound water to depths of at least 50cm on which sorghum is planted.

Within the main bund there may be smaller similar bunds which impound less runoff on which planting can be done earlier.

Micro-catchment systems

This involve runoff generation within the farmer's field and subsequent concentration on either a single crop especially fruit trees, a group of crops or row crops with alternating catchment and cropped area mainly along the contours. A number of within-field RHM systems fall under this technology, in which crop land is subdivided into micro-catchments that supply runoff either to single plants (e.g. pawpaw or oranges) for example *Negarims* in *Kitui*, Kenya or a number of plants (e.g. maize, sorghum etc) in case of c*hololo* pits in *Dodoma*, Tanzania. Pitting techniques, where shallow planting holes (< 25cm deep) are dug for concentration of surface runoff and crop residue/manure, are found in many farming systems throughout SSA. They come in many names and include *zai* pits (Burkina Faso), *matengo* pits (southern highlands of Tanzania) and *tumbukiza* for *Napier* grass and banana or pawpaw pits (Kenya). Terraces and ditches, which retain soil moisture, are other micro-catchment techniques promoted and adopted in SASE. The following are more examples:

- *Fanya juu* terraces, which are made by digging a trench along the contour, and throwing the soil upslope to form an embankment. They have made a very significant impact in reducing soil erosion in semi-arid areas with relatively steep slopes (Thomas, 1997; Tiffen *et al.*, 1994). They have been used for RHM by incorporating tied ridges in the channel with closed outlets.
- *Fanya chini*, in which the soil is thrown downslope instead of upslope, was developed in *Arusha* region, Tanzania.
- Contour bunds, e.g. stone bunds and trash-lines in dry areas of southern Kenya and retention ditches and stone terraces in Ethiopia. Yields of sorghum are reportedly increased by up to 80% using contour bunds in northwestern Somalia (Critchley, 1987).
- Micro-basins, which are roughly 1.0m long and <50cm deep, are often constructed along the retention ditches for tree planting (e.g. northern *Tigray*, Ethiopia) (Lundgren, 1993). Sweet potato ridges/bunds in southern Uganda fall under this category (Kiggundu, 2002). In *Kwale* district of Kenya, tied ridges and small basins have been reported to improve maize yields by > 70%.
- Semi-circular earth bunds (demi-lunes) are found in ASALs for both rangeland rehabilitation and for annual crops on gently sloping lands (e.g. *Baringo* and *Kitui* districts) (Thomas, 1997). Semi-circular bunds adopted for establishment of tree seedlings in denuded hilly areas in southern Uganda applies the same principle (Kiggundu, 2002).
- *Negarims* micro-catchment are regular square earthdams turned 45° from the contour to concentrate surface runoff at the lowest corner of the square (Hai, 1998) are found in Eastern Province of Kenya.
- Large trapezoidal bunds (120m between upstream wings and 40m at the base) have been tried in arid areas of *Turkana* district, northern Kenya for sorghum, trees and grass (Thomas, 1997).
- Infiltration trenches/ditches, which are dug along the contour, at specified intervals according to the slope, for retaining runoff in banana plantations in southwestern Uganda, *Mbarara* and *Rakai* districts (Kiggundu, 2002).

- Circular depressions (3-4m in diameter and <1.0m deep) where a variety of crops are inter-cropped and literally allows no runoff from the fields are practiced in southern Ethiopia.

2.1.5 Role of RHM Systems

Potential for improving food production

There are a number of promising interventions for improving water availability either for crop production or other uses in the dry parts of the SSA region. A few techniques especially for irrigation have been tested and proven successful but majority, which are mainly land users' innovations remain unproven. It is evident that the introduction of new technologies without land users' participation, however novel they may be, has not been successful. One such project is the multi-million *Bura* irrigation scheme in Kenya. On the other hand, the land users' ingenuity has certainly paid dividends. The challenge now is to evaluate land users innovations and traditional systems to determine their appropriateness in solving the recurrent food crisis in the region. Clearly the development of the SASE represents the highest potential for further economic advancement in the region. The major challenge is how to utilize the available water—the most limiting factor to economic activity in the dry areas.

Currently, most countries in the region are not able to marshal financial resources to enable bulk water transfers (e.g. inter-basin transfers), or dam and reservoir construction for most of the SASE. The pragmatic way forward is in the development of least-cost small-scale RHM systems by the communities and individuals who live within these areas. Mere survival instinct has led many land users in the SASE to improvise various indigenous runoff-farming systems. However, due to limited technical resources, these indigenous runoff-farming systems are poorly designed and operated. Therefore, a great benefit can be realized through technical improvements of the existing water harvesting initiatives. This can be accomplished by first understanding and evaluating the various systems being used in the region and comparing their performance vis-à-vis the prevailing local conditions.

As water becomes more and more scarce, there is a need for an integrated approach to water management that encompasses all water users, types of water uses and sources of water. Water management, however, can never be an aim in itself, it is an integral part of farm and land husbandry and its objective should always be to protect and improve the land users' situation (LEISA, 1998). Nevertheless, high-external-input techniques may be too expensive for smallholders or are inappropriate to local biophysical and social conditions. Many land users would benefit from low-cost techniques more suited to their conditions and needs, and which also ensure an increase in water use efficiency and conservation. It is encouraging that land users have developed many low-cost water saving techniques. Unfortunately, although most of these innovations remain unrecognized, many of them are within the reach of the land users. Therefore, according to LEISA (1998), water scarcity can be challenged!

Traditional water harvesting systems are characterized by flexibility and endurance and are strongly associated with the people who live in marginal environments. Thus different areas will have different techniques for harvesting and applying water. Although the potential for water harvesting has not been fully assessed, this potential is probably quite large in the GHA where food security is a major concern. Recently, renewed interest has been shown in water harvesting in SSA, probably as a result of increasing pressure on land, which forces more and more people into dry areas (Oweis *et al.*, 1999). This new trend could also be attributed to failure of more conventional methods and changing environments forcing people to adopt new survival strategies. Therefore, water harvesting has a high potential for improving food security and reducing over-dependency on food aid. However, for this potential to be realized, appropriate techniques need to be identified for particular areas within the region. The case studies will contribute towards identifying different techniques land users in the region have already tested and approved, and look into ways of improving the adopted technologies.

RHM is a promising technology for improving the livelihoods of many inhabitants of the vast dry regions of the world. RHM can be viable in areas with rainfall as low as 300mm yr^{-1} (Kutch, 1982). However, Pacey and Cullis (1986) gave a more conservative range of rainfall, 500-600 mm yr^{-1}. But, Kutch (1982) further stated that annual rainfall is not the most important criterion. Nevertheless, the technology has been used to sustain food production in the Negev desert of Israel with meager rainfall of about 100mm yr^{-1} (Shanan and Tadmor, 1976). Ironically, most of the famine stricken areas of Africa receives much more than 100mm yr^{-1} of rainfall.

Thus many parts of the SSA could tremendously improve food security through RHM, which aim to supply the deficit between rainfall and evapotranspiration during the growing season. In case of RHM for supplemental irrigation, the deficit is maintained by supplying water to the crops during the critical periods. Some experts regard irrigation as the only viable method of agricultural production in the ASAL (Pacey and Cullis, 1986). But history has proved otherwise especially for small scale farming systems. Therefore, promotion of RHM should take into consideration the perceived low rates of financial investments, especially in runoff farming, compared to irrigated agriculture. RHM minimizes some of the problems associated with irrigation such as competition for water between various uses and users, low water use efficiency, and environmental degradation. It is a simple, cheap and environmentally friendly technology, which can be easily managed with limited technical skills. The technology can also be integrated with many land use system, hence it is appropriate for local socio-cultural, economic and biophysical conditions. Furthermore, there are many traditional water management techniques still being used to make optimal use of available rainfall (LEISA, 1998).

Reduction of conflicts over water resources

Extensive areas of the SSA countries are not well endowed with water resources. This scarcity is aggravated by poor distribution of water resources in most countries. For instance, in Kenya, less than 20% of the country has adequate water resources for rainfed agriculture. In the vast dry areas, the main challenge is, therefore, to increase water supply through more efficient utilization of rainfall. It is evident that water scarcity is one of the main drawbacks to substantial development of the ASAL. This scarcity has led to persistent conflicts over use and access to existing water supply. The conflicts involve different water users and uses.

More often than not, different clans especially within the pastoral communities, in the ASAL have been engaged in increasing conflicts over the control and use of communal water sources and grazing land. Cross border conflicts leading to severe clashes have also occurred over control of natural resources. Notwithstanding existing traditional institutions, that to some extent have promoted peaceful coexistence, the conflicts seem to get worse by the day as water resources become scarcer. Hence, one of the logical ways to contain the situation is to provide adequate water and food supply. This approach has apparently worked well in northeastern Kenya, especially in *Wajir* district, where a local NGO has assisted in the construction of water pans to store rainwater for different clans (Githinji, 1999). This is a case where low technology—water harvesting—has proved itself, not only as a water supply system, but also as a conflict resolution mechanism. In addition, the technology has led to improved food security and living standards through provision of water for domestic, livestock and agricultural purposes. This technology has also created employment besides being easily replicable. Similar cases will be articulated in the proposed project and hence ways of dealing with the twin problem of food security and conflicts over natural resources which is prevalent in the GHA region.

Moreover, the case studies considered interrelated environmental governance and gender issues affecting food security and water availability. These include sustainable environmental management strategies and traditional institutions that are involved in the well being of the community and management of conflicts over use of natural resources. Many of the conflicts, especially inter-clan conflicts, are normally aggravated by food insecurity and competition over scarce natural resources.

Conflicts over natural resources, especially water and land, have been politicized in the SSA. According to Mathenge (2002), the issue of water is of equal importance on the political scene as security in *Laikipia* district in upper *Ewaso Ng'iro* river basin. Large-scale horticultural farming by wealthy local and international concerns on the slopes of the *Aberdares* and Mt. Kenya has depleted the mountain streams that used to be the main sources of water leading to upstream-downstream conflicts. Small-scale farmers along the streams have also contributed to water conflicts by abstracting water, in most case (more than 70%) illegally, for irrigation. During extreme dry spells, the provincial administration normally intervenes by banning water abstraction for irrigation. Otherwise downstream users would organize themselves and destroy water diversion structures upstream.

Insecurity too has contributed to the problem as many farmers have abandoned livestock rearing—attractive to cattle rustlers—to try out farming. This has increased conflicts over water rights and food insecurity. The politicians in the area have threatened to lead the affected communities to storm horticultural farms over water conflicts, while others have proposed that government impose levies on major horticultural producers to raise funds to construct and maintain reservoirs to harness flood waters. Hence RHM could play a major role in conflict resolutions, especially in drier areas of GHA. Some large-scale horticultural farmers have already adopted RHM by constructing large earthdams to harvest runoff to supplement limited water for irrigation. Another form of conflict occurs during the rainy season over limited runoff on shared road/footpath drainage. This is becoming common in *Ng'arua* division of *Laikipia* district where adjacent farmers compete and some times fight over diversion of runoff to their farms, especially those with farm ponds for storing water for use during inter-seasonal droughts—to mitigate water stress during critical growth stages. This kind of conflicts could be addressed through improved community management systems.

2.1.6 Hydrological Impacts: Limits of Up-scaling RHM Systems

Retention of (rain)water in the upper parts of a catchment by RHM systems may have negative hydrological impacts on downstream water users. Downstream access to water as a result of increased water withdrawals upstream is an issue of concern, but it is assumed that there are overall gains and synergies to be made by maximizing the efficient use of rainwater at farm level (Rockström, 2001). However, up-scaling of RHM—increasing adoption—could have hydrological impacts on river basin water resources management. The study aimed at assessing downstream-upstream interaction related to increased adoption rate—retaining more water in the watershed—in the water scarce *Ewaso Ng'iro* river basin in Kenya. Upgrading rainfed agriculture, through the adoption of RHM systems in the SASE, require proper planning of land management at river basin scale, rather than conventional focus on farm level. However, one of the challenges is to quantify the amount of water retained upstream and its impacts downstream. Hydrological variations at different spatial scales even make this assessment more complex. For example, deep percolation or runoff from one farmer's field may be considered as water loss, but it would be beneficial to downstream land users.

In the past, runoff has been as being destructive and needed to be diverted from agricultural lands as witnessed by over 30 years of soil conservation practices in Kenya. However, radical transformations are required, where surface runoff from upstream watershed entering a farm will no longer be seen as a threat to be disposed of or diverted away, but as a resource to be harnessed and utilized to improve rainfed agriculture. Such transformations are complex, especially among small-scale farmers, since even runoff from a small catchment will involve multiple land users. Presently there is little attention given to ownership and management of locally produced runoff, but this is expected to become a paramount issue if runoff is to be optimally managed on a larger scale for productive purposes. In *Laikipia* district of Kenya, conflict over runoff diversion and utilization for crop production is a reality (Kihara, 2002). The situation may become much worse with the growing realization of the benefits of RHM systems, especially for resource-poor small-scale farmers, who depend solely on rainfed agriculture. Already even with limited RHM systems, downstream-upstream conflicts between pastoralists and farmers (who divert meager stream water for irrigation) are very common particularly during the dry periods. Even farmers are also being involved in conflicts, among each other and with the downstream land users—both farmers and pastoralists in this water-scarce river basin. The Indian experience on communal rainwater management (Agarwal *et al.,* 2001; Agarwal and Narain, 1997) may provide useful background in an attempt to develop sustainable RHM up-scaling strategies in SSA. The Indian experience is based on the principles of integrated water resources management (IWRM), whose adoption would enhance sustainable up-scaling of RHM systems at different spatial scales. However, one of the challenges is adoption of IWRM and interpretation & extrapolation of hydrological processes at different spatial scales. For example, the effects of small-scale RHM systems (farm level) on overall river basin stream flows.

In the view of ensuing competition and conflicts over limited water resources, the hydrological impacts of up-scaling RHM in a watershed/river basin need urgent attention. The Indian experience, where the *Rajastan* Irrigation Department, fearing that the communal RHM systems would threaten water supply located downstream ordered its destruction (Agarwal *et al.*, 2001), indicates the need for policies, legislation and new institutional order to manage adoption of RHM systems, especially for agriculture and livestock. In Kenya, and many countries in SSA, such policies and legislation are unfortunately lacking or inadequate. This poses another challenge in the assessment of hydrological impacts of up-scaling RHM systems. Therefore, there is a need for research to provide data to assist decision and policy makers formulate sustainable river basin water resources management strategies.

Several hydrological studies at watershed and river basin scales have shown that upstream shifts in water flow partitioning may result in complex and unexpected downstream effects, both negative and positive, in terms of water quality and quantity (Vertessey *et al.*, 1996). In general though, increasing the residence time of runoff flows in a watershed, e.g., by RHM systems may have positive environmental as well as hydrological impacts downstream (Rockström *et al.*, 2001). The hydrological impacts at watershed/river basin level of up-scaling RHM systems are still unknown and require further research. This Ph.D. study aimed to shed some light on this issue.

Increased water retention and withdrawals in rainfed and irrigated agriculture may have negative implications on water availability to sustain hydro-ecological ecosystem services. The expected shifts in water flows in the water balance would affect both nature and economic sectors depending on direct water withdrawals (Rockström *et al.*, 2001). Upgrading rainfed agriculture through RHM that enables dry spells mitigation, would involve the addition of water, through storage of runoff, to the rainfed system. The cumulative effect of RHM may have an impact on downstream water availability within a river basin scale. The effects are bound to be site specific and need to be studied further (Rockström *et al.*, 2001).

The potential of developing farm ponds and earthdams for supplemental irrigation in SSA, is determined by a set of site specific biophysical and socio-economic factors (Rockström, 1999), which include practiced farming systems, population pressure, formal and informal institutions, land tenure, economic environment and social structures. Thus hydrological impacts cannot be assessed in isolation. It is important to analyze the downstream effects on water availability, for example, for domestic, agricultural and livestock needs, as well as health and environmental impacts, before introducing a technology which retains water upstream, and possibly reduce river flows. Therefore, the study was expected to enhance knowledge required to develop strategies for enhancing sustainable water resources management in water-scarce river basins, in general SSA, and in particular Kenya. The water-scarce river basins are, besides ensuing water crisis, experiencing land use changes that further complicate the situation. Thus the current challenge is how to assess hydrological impacts of land use changes on water resources development and socio-economic development in water-scarce river basins, which are occupied by diverse water users.

2.2 Description of Study Area

2.2.1 Ewaso Ng'iro river basin

The *Ewaso Ng'iro* is the largest out of the five major river basins that make up the Kenyan drainage network, and covers 210,226 km². The study was based on the upper *Ewaso Ng'iro* basin, which constitutes the upper stream section of this drainage area, covering 15,251 km^2. It is part of the *Juba* basin, which covers parts of Ethiopia, Kenya and Somalia (see Fig. 2.3). It is situated between latitudes 0^0 20' south and 1^0 15' north and longitudes 35^0 10' east and 38^0 00' east.

The river basin drains from the Rift Valley escarpment to the west, *Nyandarua* (formerly *Aberdare*) ranges to the southwest, Mt. Kenya to the south, *Nyambene* hills to the east, *Mathews* range to the north while the downstream outlet is at *Archer's Post*. The upper *Ewaso Ng'iro* basin traverses the administrative districts of *Laikipia, Nyeri, Nyandarua, Meru, Nyambene, Isiolo and Samburu. Laikipia* district occupies about 50 % of the entire upper *Ewaso Ng'iro* basin. Although the main *Ewaso Ng'iro* river originates from the *Nyandarua* ranges, most of the flow comes from the tributaries that drain Mt Kenya catchment (Decurtins, 1992). Whereas the surface flows from *Ewaso Ng'iro* river drains into the *Lorian* swamp in north-eastern, Kenya, subsurface flow continues eastwards to recharge rivers in Somalia, which eventually drain into the Indian Ocean.

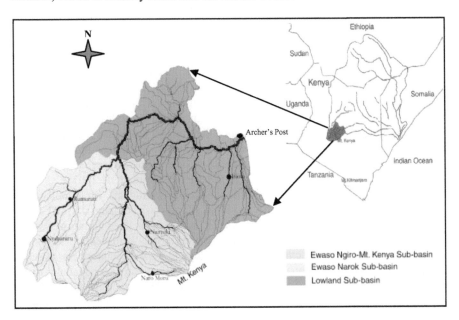

Fig. 2.3. Location and main sub-basins of *Ewaso Ng'iro* river basin

Topography, soils and vegetation

The topography is dominated by Mt. Kenya and the *Nyandarua* ranges to the south and the *Nyambene* hills to the east of the basin. Altitude ranges from 862m at *Archer's Post* to about 5200m at the summit of Mt. Kenya (Jaetzold and Schmidt, 1983). The upper mountain slopes are undulating to rolling slopes (5-16%) with deeply incised V-shaped valleys, where elevation ranges from about 2,500 to over 4,000m, while the lower parts consist of broad ridges dissected by rivers and streams. The extensive gently undulating *Laikipia* plateau at an elevation of 1,700-1,800m occupies most of the central region. The north and north eastern parts of the basin is characterised by undulating to rolling topography comprising of uplands, hills and minor scarps with many of the latter having outcrops of the basement complex rocks at their tops. Here altitudes drop rapidly from about 1,700m to less than 1000m towards the extreme north-eastern region.

The upper *Ewaso Ng'iro* river basin is a highland-lowland system and changes in elevation gives rise to a dramatic climatic and ecological gradient, from humid moorlands and forests on the slopes to arid *Acacia* bushland in the lowlands, with a diverse pattern of land use (Decurtins, 1992). The natural resources of this highland-lowland system are under pressure in the highland (resource rich) due to dynamic land use changes and intensification resulting in resource degradation and in the lowland (resource poor) due to immigration, accompanied by inappropriate land management practices, and marginalization of the indigenous community. The upper *Ewaso Ng'iro* basin can be divided into three zones depending on topography: mountain slopes, lower highlands and lowlands.

The mountain slope is the forest zone of Mt. Kenya and the *Nyandarua* ranges. The natural vegetation is evergreen mountain forest. Ground cover consists of perennial herbs and a thick layer of litter. The lower parts of the natural forest have been cleared and replaced by forest plantations mainly of cypress and pine, large-scale barley and wheat farming and small-scale farming mainly growing potatoes and horticultural crops. The major soil types in the mountain slopes are stony *mollic Cambisols* and *mollic Andosols* of medium depth in the upper slopes of Mt. Kenya while in the middle parts are deep *mollic Andosols* and in the lower parts are deep to very deep *humic Alisols* (Mbuvi and Kironchi, 1994).

The lower highlands (foot slopes) constitute the area adjacent to the lower mountain slopes and the immediate *Laikipia* plateau. The landforms are at elevations of between 1,800 and 2,100m. The slopes ranges from nearly flat (1-2%) on the ridge tops to 8% in valleys. The soils of the foot slopes and plateau are underlain by *Phonolites* from Mt. Kenya volcanic activities and those on the ridges are well drained, deep, dark red to dark brown friable clay, *Luvisols* and *Phaeozems*, while those on the flat areas are imperfectly drained, deep grey to black firm clay, *Vertisols* and *Planosols* (Mbuvi and Kironchi, 1994). The area has been subdivided into small plots ranging from 1-5 ha and is settled by small-scale farmers growing mainly maize and beans as food crops and wheat as a cash crop. Most farmers keep cattle, sheep and goats, which are grazed in small patches of uncultivated land or in the cultivated plots after the crops are harvested.

The lowlands (basement area) comprise of landforms with slopes ranging from 2-8% at an elevation of between 1600 and 1800m. It is covered by metamorphic rocks consisting mainly of *gneisses* and *migmatites* with a few granite outcrops. Soils are well drained to excessively drained, dark reddish brown in colour and range from shallow to deep. Vegetation consists of open dry thorn bushland dominated by *Acacia* trees. The area is mainly used for extensive livestock grazing and as a habitat for wild animals. There are private large-scale ranches and communal grazing by the *Maasai* pastoralists.

Climate and agro-climatic zones

The elevation gradient, which determine the rainfall pattern give rise to various climatic zones ranging from humid to arid (Fig. 2.4). The long term rainfall analysis shows high spatial and temporal variation ranging from 300-2000 mm yr^{-1}, with a mean of about 700 mm yr^{-1}. Rainfall pattern indicates a recurrence of wet-dry cycles of 5-8 years (Gichuki, 2002). Rainfall variation affects river discharge, which varies from 0-1,627 m^{3}s^{-1} at *Archers' Post* over the period 1960-2003. Rainfall intensities are usually high averaging about 20-40 mm hr^{-1} while higher intensity storms of up to 96 mm hr^{-1} have been recorded (Liniger, 1991). There are three main rainfall seasons, namely long rains (March - June), continental rains

(July - September), short rains (October - December). The long rains and short rains contribute 30-40% and 50-60% of annual rainfall respectively.

Rainfall distribution (Fig. 2.4) and agro-climatic zones are primarily determined by topography. As a result, the basin has a wide climatic range from agro-climatic zone I-IV (Sombroek *et al.*, 1980). Table 2.1 shows the characteristics of agro-climatic zones and their distribution in Kenya. Annual rainfall in the mountain zone increases with altitude up to around 3,000-3,500m. The climate is semi-humid and area falls under agro-climatic zone III. Rainfall in the foot slopes (lower highlands) is very variable ranging from as low as 300mm yr^{-1} to a maximum of about 1000mm yr^{-1} with the mean around 700mm yr^{-1}. The climate is semiarid to semi–humid and it falls under agro-climatic zone IV.

Table 2.1. Characteristics of agro-climatic zones and their distribution in Kenya

Zone	P/E_o ratio	Designation	Area $* 10^3$ (km^2)	% of Area
I	> 0.80	Humid	254	4.3
II	0.65-0.08	Sub- humid	238	4.1
III	0.50-0.65	Semi-humid	257	4.4
IV	0.40-0.50	Semi-humid to semi-arid	287	4.9
V	0.25-0.40	Semi-arid	873	15.0
VI	0.15-0.25	Arid	1,264	21.7
VII	< 0.15	Very arid	2,653	45.6

Source: Sombroek *et al.* (1980)

Note: P is mean rainfall (mm yr^{-1}), E_o is mean potential evaporation (mm yr^{-1}) and P/E_o is moisture availability ratio.

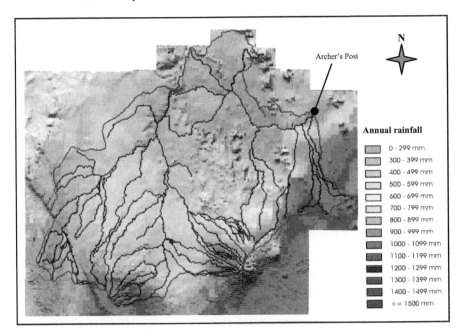

Fig. 2.4. Location of *Ewaso Ng'iro* river basin showing mean annual rainfall distribution

In the lowland, rainfall amount ranges from about 250-600mm yr^{-1} with the long term mean around 450mm yr^{-1}. The lowland has a bimodal rainfall distribution with the long rains coming between March and May and the short rains in October and November. Mean potential evaporation ranges from 2,000-2,300mm yr^{-1}. The climate is generally arid and falls in agro-climatic zone V. The annual rainfall amounts are highly variable, tapering from over 2,000mm yr^{-1} on the *Nyandarua* ranges to below 365mm yr^{-1} in the drier northern and eastern areas.

The upper *Ewaso Ng'iro* basin is partly located on the leeward slopes of Mt. Kenya and the windward slopes of the *Nyandarua* range, which drastically affect its climate. However, despite the relatively high amount, rainfall distribution is such that the seasonal amounts are insufficient for proper crop growth in most parts of the basin. The eastern region has a clear bimodal distribution with highest rainfall in April and October (Berger, 1989). In the western and north-western regions, besides the bimodal pattern, continental rainfall falls between June and September. The central region is a transition zone where the two patterns overlaps and hence it is the driest region. Rainfall intensities are usually high averaging about 20-40mm hr^{-1} and higher intensity storm of up to 96mm hr^{-1} have been recorded (Liniger, 1991). Due to the difference in altitude, mean annual temperatures in the river basin range from below 10^{0}C at the top of Mt. Kenya to over 24^{0}C at *Archer's Post*. In comparison to other ASALs of Kenya, temperatures are relatively low, with mean annual values of 18-20^{0}C. This is due to the higher altitudes and the effects of cool winds that descend from the mountains.

Water resources

The main sources of water in the basin include rivers, groundwater, dams and water pans. The main tributaries of *Ewaso Ng'iro* river are divided into two sub-catchments: *Nyandarua* ranges and Mt. Kenya sub-catchments. Mt. Kenya catchment is far more important for the discharge of *Ewaso Ng'iro* river than the *Nyandarua* sub-catchment. Seven rivers originate from the northern slopes of *Nyandarua* ranges, i.e. *Moyo, Ewaso Ng'iro, Ngobit, Suguroi, Mutara, Pesi* and *Ewaso Narok*. The rivers draining the northern and western slopes of Mt. Kenya are *Timau, Teleswani, Sirimon, Kongoni, Ontulili, Likii, Nanyuki, Rongai, Burguret* and *Naro Moru*. More than 60% of the basin inhabitants depend on the waters of these rivers for irrigation (IRIN, 2002).

Most parts of the river basin have water as the most limiting factor in agriculture and industrial development (Kithinji and Liniger, 1991). Due to water scarcity, crop production and livestock keeping which are the major land use activities are greatly constrained. Recurrent crop failure, low biomass production in the rangelands, soil erosion and high runoff losses highlight the importance of soil, water and vegetation resource management. The availability of water varies considerably through out the year. Due to rainfall characteristics water deficit increases drastically as one move from the peak of Mt Kenya (see Fig. 2.5). Excess water in the rainy season is followed by shortage in the subsequent dry season.

The situation continues to deteriorate as the small-scale farmers continue to settle within the basin. The need of the various land users who depend on *Ewaso Ng'iro* river differs. The pastoralists in the lower reaches uses the water to cater for their domestic needs and to water their animals, while small-scale farmers in the upper regions in addition use the water to irrigate the crops. The main water users are not the pastoralists living in the lower reaches of the basin but the increasing small-scale farmers upstream. Increasing water demand in the upper river catchment is one of the causes of water conflicts between upstream farmers and downstream pastoralists. Most of the river-based water supply systems have no storage facilities to bridge the dry season. As a result, 60-95% of the available river water is abstracted during the dry seasons in the upper reaches, with up to 90% being unauthorized (Wiesmann *et al.*, 2000).

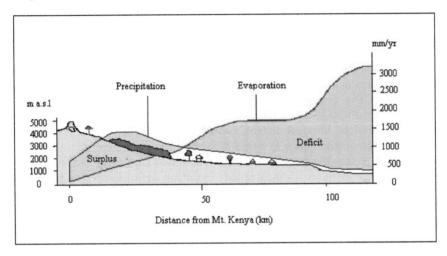

Fig. 2.5. Variation of annual water deficit with altitude and distance from Mt. Kenya

Population and land use

Population has increased from 50,000 in 1960 to 500,000 in 2000 (Wiesman *et al.*, 2000). The growth rate is estimated at 5-6% per annum (GoK, 1999) and is a result of natural increases and immigration from adjacent densely populated areas. Population in upper *Ewaso Ng'iro* river basin averages about 60 persons per km^2, but the distribution ranges from 212 persons per km^2 on the highland small-scale farming areas, to less than 24 persons per km^2 in the plateau areas (Huber and Opondo, 1995). The diverse population densities relate to different land use systems. Population pressure induces land use changes, which are accompanied by a diversity of soil and water management techniques (Kironchi, 1998). The area under cultivation on foot and savannah zones had more than tripled over the last 20 years, while water abstractions had dramatically increased (IRIN, 2002).

Change in elevation gives rise to a dramatic climatic and ecological gradient, from humid moorlands and forests on the slopes to arid acacia bushland in the lowlands, with a diverse pattern of land use (see Fig. 2.6). Large-scale commercial ranches for beef and dairy cattle, and wildlife conservation take up a large proportion of the basin area. The drier northern regions of *Mukogondo, Isiolo* and *Wamba* are occupied by communal grazing lands and group ranches used by the pastoral communities who include the *Maasai, Samburu, Boran, Turkana, Ndorobo* and *Somalis* (Thurlow and Herlocker, 1993). Most of the river basin is comprised of arid and semi arid lands. Despite the marginal status, a significant proportion of this area has been subdivided and small-scale farming established in the former white "highlands".

Ecological constraints of the SASE are numerous. These include low carrying capacity due to low and variable primary production resources and large variability in the rainfall amount and distribution. The inevitable land use changes put pressure on the fragile environment, resulting in a problem on how to sustain production while at the same time conserving the natural resources. Conflicts in the various traditional land use activities have been experienced and the natural ecological balances within the individual locations have been threatened. The changes are in some cases accompanied by environmental degradation, declining primary production resources and impoverishment making the management and planning for sustainable use of natural resources extremely difficult.

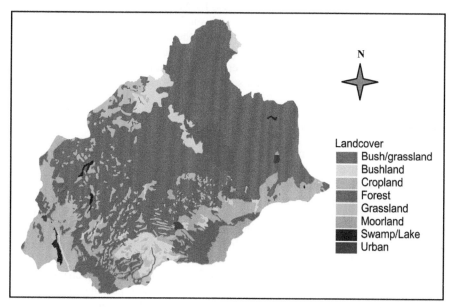

Fig. 2.6. Land cover which relates to diverse land uses in upper *Ewaso Ng'iro* rive basin

Two game reserves, *Samburu* and *Buffalo* are found within the basin and reserved exclusively for wildlife. There are a number of privately owned wildlife sanctuaries found in most parts of the basin, and often cause human-wildlife-livestock conflicts over water, land and pasture. Cultivated land occupies a small proportion of the basin. Forests are found on the higher altitudes of Mt. Kenya, *Nyandarua* and the *Nyambene* hills. Forests having less canopy and bushland are found on the *Mukogondo* and *Engare* highlands, *Mathews* ranges, hills and minor scarps.

There are contrasts in the livestock management between the large-scale commercial ranches and the communal grazing areas, where overgrazing occurs and land degradation is quite evident. Tourism is an important economic activity, the main attraction being wildlife. There are different wildlife management practices which include trust land under the county council, privately owned game ranches, gazetted government forests and communal grazing lands. The animal population and diversity of species change with each management system.

The diversity in land use and land management practices ranges from highly mechanized and modern farming techniques on the large-scale farms and vegetable/green houses irrigated farms (e.g. in *Timau* and *Naro Moru* areas), to the manual cultivation with low technological and material inputs on the adjacent small scale farms. Large-scale wheat and barley farms are found in some parts around Mt. Kenya and *Nyandarua* ranges. Adjacent to these are small-scale settlements under mixed farming systems. Here maize is the predominant crop, supplemented by beans, peas, potatoes and vegetables (Kohler, 1987).

Due to hydro-ecological constraints, many farmers (small-and large-scale) are adopting various RHM systems to improve agricultural production. Some of the RHM technologies are on-farm water pans/ponds and earthdams for supplemental irrigation and livestock, conservation tillage, soil and water conservation, road drainage runoff harvesting and roof catchment. With continued pressure on land and inadequate rainfall and water resources, RHM systems seem to be one of the options for improving food production and socio-economic development in the basin. The *Ewaso Ng'iro* river basin epitomizes the increasingly common situation in Kenya of population pressure resulting in excessive abstraction of river water in the humid highlands leaving little water for downstream users in SASE (ADF, 2005). Hence solutions to pressure on scarce natural resources in this basin would have a wider application.

2.2.2 Naro Moru Sub-basin

The *Naro Moru* river basin was used as a case study to assess the hydrological impacts of land use changes in *Ewaso Ng'iro* river basin. The *Naro Moru* river basin, which has is173 km^2, spreads westwards from the peak of Mt. Kenya to the semi-arid *Laikipia* plateau between latitudes 0° 03' and 0° 11' south and longitudes 36° 55' and 37° 15' east. The altitude of *Naro Moru* drainage basin ranges from about 5200m at the peak of Mt. Kenya to 1800m at its confluence with *Ewaso Ng'iro* river. The river sub-basin lies within humid to semi-arid agro-climatic zones (see Fig. 2.7), and is characterized by low amount of rainfall due to its location on the leeward side of Mt. Kenya (see Fig. 2.3). The catchment rapidly changes from highland to lowland climatic conditions with the smaller part having a humid climate exhibiting surplus rainfall and the larger part being a semi-arid environment.

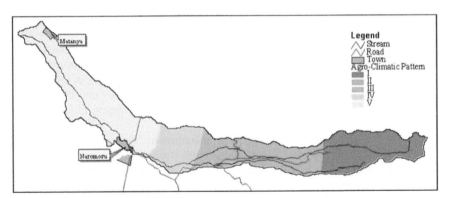

Fig. 2.7. Agro-climatic zones and rainfall distribution in *Naro Moru* sub-basin

The rainfall distribution is bimodal and highly skewed with the highest, rainfall amounts and intensity, being received in the upper forest zone (Gichuki *et al.,* 1998). Mean rainfall within the catchment increases from 650mm yr^{-1} at the outlet to 1500mm yr^{-1} at 3300m altitude and then drops to 500mm yr^{-1} in the moorland (NRM, 2003). More than 70% of the catchment receives on average < 900mm yr^{-1}. The potential evaporation is high, on average above 2500mm yr^{-1} and exceeds annual rainfall for most part of the year. The deficit increases towards the savannah zone. Although *Naro Moru* catchment accounts for only 1.1% of the upper *Ewaso Ng'iro* basin, it can be taken as a representative sub-basin, which reflects typical constraints and challenges affecting socio-economic development in the entire basin. These include unreliable rainfall patterns and quantities, wide variety of stakeholders with diverse interests, highland-lowland interdependence, unequally distributed natural resources, and decreasing river flows, which lead to conflicts between upstream and downstream users (Aeschbacher *et al.*, 2005).

2.3 Methodology Overview

The research methodology was based on achievement of the study objectives and broadly addressed the following: review and analysis of RHM system; selection of specific study sites; development of conceptual and analytical framework; spatial runoff monitoring; agro-hydrological evaluation of RHM systems; hydro-economic evaluation of RHM systems; and assessment of hydrological impacts of RHM on river flows. The methodology is based on the conceptual and analytical framework (*Chapter 3*) developed to assess the hydrological impacts of land use changes aimed at upgrading rainfed agriculture (e.g. RHM systems and irrigation) and improving rural livelihoods.

2.3.1 Review and Analysis of RHM Systems

Extensive field evaluation of existing RHM systems was carried out in *Laikipia* district and their distribution used to select specific study sites for detailed analysis. *Laikipia* district was one of the six case studies carried out for comparative analysis of the potential of RHM systems in the GHA. Different RHM systems were evaluated and documented (Ngigi, 2003a). The distribution of the RHM systems and sites selected for the study are presented in Fig. 2.8. The results formed the basis of assessing the hydrological impacts of RHM on river basin water management. The selection was also based on adoption rate and/or potential for adopting different RHM systems. Three study sites were selected for detailed agro-hydrological and hydro-economic analysis: (a) semi-humid *Muramati* (near *Kalalu*); (b) semi-humid to semi-arid *Matanya*; and (c) semi-arid *Mutirithia* (near *Naibor*) as shown in Fig. 2.8.

2.3.2 Spatial Runoff Monitoring

Runoff monitoring was conducted at different spatial scales; runoff plot, field/farm, medium catchment, sub-basin and basin scales as shown in Fig. 2.9. At field scale, runoff from both in-situ (conservation tillage) and on-farm storage system for supplemental irrigation were monitored and compared with traditional farming systems—without RHM systems. Field monitoring was also supplemented by data from previous studies. For sub-basin and basin scale, available long term river flow records were used[5]. Whilst field scale refers to farm level, medium catchment scale refers to watershed for medium-large scale rainwater storage structures such as earthdams and water pans mainly for livestock water supply, but can also be used for irrigation and domestic water supplies.

---

[5] At the river basin scale, long term records (1960-2002) at the upper *Ewaso Ng'iro* river basin outlet (*Archer's Post*) were used. Whilst at sub-basin scale 4 river gauging stations of *Naro Moru* river basin, with records spanning from 1983-2002 were used.

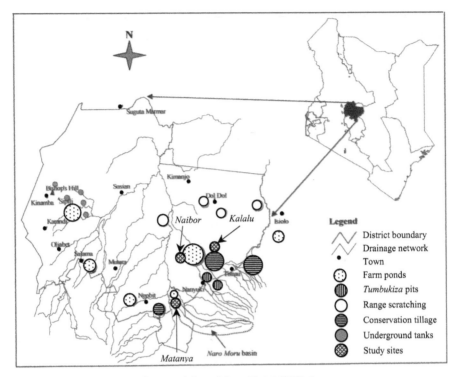

Source: Adapted from Ngigi (2003a)

Fig. 2.8. Location of different RHM systems that guided selection of study sites in *Laikipia*

Plate 2.1 shows experimental set-up for runoff monitoring at field and medium catchment scales. The field scale sites instrumentation included a manual rain gauge, evaporation pan (class A pan), pipe sampler, graduated staff and on-farm storage pond. Due to the cost implication, four medium catchment study sites, which were within the *Naro Moru* sub-basin, were not all fitted with runoff monitoring equipments. Instead, loose runoff monitoring for a number of rainfall events were conducted. This involved either use of road crossing culverts (in *Matanya* and *Thome*), natural or gully outlets (in *Acacia* and *Kieni*) to monitor runoff. Loose monitoring involved using a float, stop watch, current meter, measuring tape and graduated staff. Runoff flow was estimated from the velocity and cross-section area at the outlet point.

Plate 2.1. Layout of study sites for runoff monitoring at field and medium catchment scales

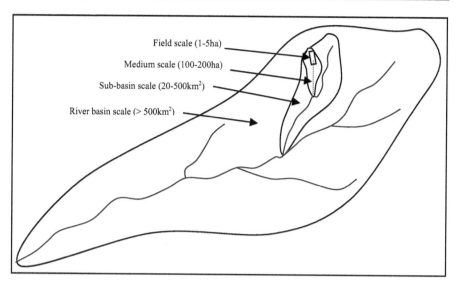

Fig. 2.9. Spatial hydrological scales for assessing the impacts of RHM in a river basin

2.3.3 Agro-hydrological Evaluation of RHM Systems

Agro-hydrological evaluation of on-farm storage system for supplemental irrigation was accomplished through water balance analysis. The data for water balance analysis were measured and amount of runoff available for crop production estimated. Seepage losses were very high and ultra-violet resistance plastic lining was used to control it on some of the farm ponds (*details in Chapter 4*). Water use for crop production was monitored using 20-litre low-head drip irrigation systems. Farmers recorded the amount of water applied to kitchen gardens planted with a variety of vegetables. The timing of supplemental irrigation was determined from soil moisture balance and dry spell analysis. Similar experimental setup was used for monitoring runoff on farms with and without conservation tillage (*details in Chapter 5*). Maize biomass and grain yields were also measured.

2.3.4 Hydro-economic Evaluation of RHM Systems

To assess the potential and willingness of farmers to invest in RHM systems, hydro-economic evaluation was carried out. It entailed hydrological risks related to rainfed agriculture and economic assessment of improving agricultural production through adoption of RHM systems. Hydrological risks evaluation was based on water balance analysis. The hydrological risks are related to inadequate and poorly distributed rainfall, unpredictability of rainfall onset and occurrence of intra-and off-season dry spells. To show the viability of RHM systems for supplemental irrigation, economic and comparative analysis for with and without RHM scenario was carried out (*details in Chapter 6*).

36

2.3.5 Assessment of Hydrological Impacts of RHM on River Flows

The assessment of hydrological impacts of RHM on river flows was based on quantification of how much rainwater (runoff) is retained on agricultural land due to adoption of RHM systems. The proportion of runoff retained was determined from rainfall-runoff relationships from farms with and those without RHM systems. However, only an indication of the proportion of rainwater retained was accomplished. More quantitative assessment will need to done to predict future trend with increasing adoption of RHM systems. Nevertheless, through monitoring of historical trend of water abstractions, it is clear that RHM can reduce dry season's irrigation water abstractions through flood storage, without significant impacts on river flows (*details in Chapter 7*). The analysis was based on long term daily flow records for *Naro Moru* sub-basin and *Ewaso Ng'iro* river basin (1960-2002).

Chapter 3

3.0 Conceptual and Analytical Framework [6]

3.1 Overview

Land use changes, especially intensification of rainfed agriculture, are driven by the need to improve agricultural production and livelihoods. One such land use change is adoption of RHM systems, which aim to enhance soil moisture and runoff storage for food production. RHM systems retain additional runoff on agricultural lands which may reduce river flows for downstream users and lead to negative hydrological, socio-economic and environmental impacts in a river basin. On the other hand, rainwater storage systems may lead to positive impacts by reducing water abstractions for irrigation during the dry period. However, depending on the level of up-scaling, RHM may lead to water scarcity and related conflicts among different users and uses. Therefore, whilst improving agricultural productivity is a priority to feed the growing populations, the ensuing water crisis is becoming a nightmare and a real policy challenge in many countries. The chapter presents a people-water-ecosystem nexus based conceptual framework for assessing the impacts of land use changes in upper *Ewaso Ng'iro* river basin in Kenya. The conceptual framework presents the key issues, their interactions and how they can be addressed to enhance integrated water resources management. It assesses the hydrological impacts of land use changes on hydrological regime in a river basin. Hydrological scenarios based anticipated impacts of land use changes are presented. The results will enhance formulation of sustainable land and water resources management policies for water-scarce river basins.

Population growth induced agricultural intensification is taking place at an unprecedented rate in parts of *Ewaso Ng'iro* river basin. In semi-arid environment, where water is a major constraint to agricultural production, RHM systems are increasing in popularity (Ngigi, 2003a). The water retained by RHM systems is part of the surface water that drains to lower reaches of the river to meet downstream water requirements. Sustainable agricultural intensification dependent on RHM requires that we address the following questions: (i) How much water can be retained by RHM systems without adversely affecting the hydrologic regime, socio-economic and environmental conditions further downstream? (ii) How much would RHM systems reduce dry season irrigation demands and river water abstractions?, and (iii) What proportion of the water retained in the catchment by RHM systems is used to recharge groundwater resources and sustain dry season river flows? The challenge is to identify appropriate responses to the threats of human activities on natural hydrological and ecological regimes in river basins (Sivapalan *et al.*, 2003).

---

[6] *Based on:* Ngigi, S.N., H.H.G. Savenije and F.N. Gichuki. 2006 *(forthcoming)*. Land use changes and hydrological impacts related to up-scaling of rainwater harvesting and management in upper *Ewaso Ng'iro* river basin, Kenya. *Land Use Policy (Article in press, accepted on 28 October 2005)*

There is growing consensus for a need to improve agricultural productivity and water resources management to meet new challenges posed by increasing demand and diminishing water supply. However, the options, processes and impacts of desired change are less clear (Hajkowicz *et al.*, 2003). Thus stakeholders are searching for a conceptual framework that can integrate policy, water users' aspirations and strategic actions to achieve the desired change. Hoekstra (1998) stated that the problem in integrated water resources assessment is not a lack of appropriate tools in any of the related sectors, but rather the lack of integration of these tools and the difficulty of translating analytical results into policy-relevant information. We need to support this statement by highlighting (i) the available tools, (ii) lack of integration of these tools and (iii) difficulties in translating results into policy-relevant information. To address the most policy and management issues as perceived by users under different biophysical and socio-economic environments and taking into account needs for sustainable development, water related physical (hydrological, climatological, ecological) and non-physical (technical, sociological, economics, administrative, law) observations are a prerequisite (UNESCO, 2005).

The chapter addresses this by developing a conceptual framework for assessing hydrological impacts of up-scaling RHM in a river basin. Up-scaling here refers to both moving from smaller to larger or improved systems (vertical up-scaling) and replication of the same systems (horizontal up-scaling or scaling out, i.e. increased adoption). The conceptual framework can be used to investigate hydrological, socio-economic and environmental impacts of intensifying agricultural production through a combination of in-situ, micro- and macro-catchment RHM systems. The main focus is hydrological impacts related to up-scaling of RHM technologies and increasing demand and water abstraction for irrigation. The case of *Naro Moru* river sub-basin is used to highlight the impacts of land use change on river flows. Despite the anticipated negative hydrological impacts, RHM may reduce river water abstractions by storing water during the rainy seasons especially for irrigating crops during the dry seasons. A major challenge to promote RHM is to formulate effective land use policies and strategies which will enhance integrated water resources management at a river basin scale. It is envisaged that the conceptual framework would enhance formulation of responsive land use policies and enhance integrated water resources management in water-scarce river basins.

3.2 Situational Analysis

3.2.1 Current Situation

The land use changes have put pressure on the fragile environment, resulting in a dilemma on how to sustain production while at the same time conserving natural resources and managing upstream-downstream water conflicting (Liniger *et al.*, 2005). The conflicts can be considered at different spatial scales: between farmers at different agro-ecological zones; between farmers and pastoralists; between sedentary and nomadic pastoralists and between farmers/pastoralists and wildlife. Land use changes have been accompanied by reduction in river flows, environmental degradation and declining agricultural production. Recent land use changes have also been associated with incidences of flash floods previously unknown due to increased runoff upstream (IRIN, 2002).

The main socio-economic activity is small-scale farming (mainly rainfed agriculture and irrigation along river courses), commercial medium to large scale horticultural production for export mainly under irrigation on the semi-humid highlands. Large-scale commercial ranches for beef and dairy cattle, and wildlife conservation take up a large proportion of the basin on the vast semi-arid lowlands. The small-scale farmers mainly cultivate food crops such as maize, beans, potatoes and wheat. Vegetables are also grown under micro-irrigation for domestic consumption, local market and export—by out-growers for the large commercial farms. Most small-scale farmers keep cattle, sheep and goats, which are grazed in small patches of uncultivated land or unsettled adjacent lands.

Poor distribution of water is the most limiting factor to socio-economic development in the river basin (Kithinji and Liniger, 1991). Excess water during the rainy season is followed by severe during subsequent dry season. The situation aggravated by unsustainable land use changes. The river flows has progressively decreased by about 30% since 1960 mainly due to increasing water abstraction upstream and drought cycles, since there is no corresponding decline in rainfall trend. Water abstraction has increased from 20% in the wet season to over 70% in the dry season (Aeschbacher et al., 2005).

Water scarcity, particularly in the lower reaches of major rivers, has increased over the years and has resulted to conflicts between upstream and downstream water users (FAO, 2003; Gichuki, 2002). However, adoption RHM systems can reduce water abstractions and related conflicts. For instance, harvesting 30% of runoff in semi-arid parts of the basin can triple the available water supply (Gichuki *et al.,* 1998). This potential if exploited would minimize dry season water demands and river abstractions. However, it remains to be seen if RHM can reduce dry season water abstractions. Some of the RHM systems are farm ponds for micro-irrigation and water pans and earthdams for livestock, in-situ rainwater conservation and diversion and storage of road drainage runoff (Ngigi, 2003a). There is a growing realization that RHM can improve food production and livelihoods in water-scarce river basins. Despite the anticipated socio-economic impacts, up-scaling of RHM, may beyond a certain limit, leads to hydrological and environmental impacts (Ngigi, 2003b).

3.2.2 Anticipated Land Use Changes and Water Crisis

The anticipated land use changes and water crisis can be attributed to increasing farming activities in water deficit areas where rainfed agriculture is not sustainable. The farmers are forced by harsh climatic conditions to improve their livelihoods through intensification of rainfed agriculture, in particular adoption of RHM and irrigation to increase crop yields or stabilize yields that are normally affected by low and poorly distributed rainfall. This means increased retention of runoff on agricultural lands, which may reduce river flows during the rainy seasons. Water abstractions for irrigation also reduce river flows during the dry periods. The natural environment, and the biodiversity it contains, is threatened by both water withdrawals and water pollution (UNESCO, 2005).

The winners-losers and opportunities-constraints analysis show that upstream farmers stand to gain if they are allowed to continue retaining and abstracting more water for agricultural production. However, this may deny downstream users their source of livelihood and affect sustainability of natural ecosystems. Balancing these conflicting needs is delicate, but the conceptual framework would enhance understanding of the complexities involved and assist in developing sustainable solutions. Some of the viable options are reduction of water retention upstream, reducing water demands and construction of storage and flow regulating reservoirs.

The options for hydrologically advantaged upstream users to ensure minimum allowable flows downstream are RHM, improving water use efficiency, shifting from full irrigation to deficit and/or supplemental irrigation, reducing cropped area and shifting to lower water consumption crops. This may be expecting too much from the winners who are used to enjoying their hydrological advantage.

The disadvantaged downstream users could address water scarcity by construction of storage structures, improving water use efficiency, conjunctive use of groundwater and surface water, and demand management-oriented systems. Again, this may be expecting too much from people who have generally low incomes and are unable to invest in such measures. If no action is taken, either they will be destined to continue suffering or they will fight for water because they believe their counterparts upstream are reaping benefits at their expense. Thus, the government should intervene by formulating and implementing responsive policies and legislation to balance water demands of upstream and downstream users. Any sustainable intervention should consider the interests of the conflicting water users and other stakeholders. Therefore, understanding the people-water-ecosystem nexus form the basis of developing the conceptual framework and hence effective water resources management strategies. The policy formulation process should facilitate dialogue among different stakeholders. The current demographic, socio-economic, institutional and technological transition and ensuing land use changes means that a good knowledge base is required to inform the policy and decision-making processes. The conceptual and analytical framework attempts to provide the required information and interactions.

3.3 Conceptual and Analytical Framework

The conceptual framework is designed to assist water users, researchers and policy-makers to address complex problems of natural resources planning at the river basin scale (Hajkowicz *et al.*, 2003). A conceptual framework, based on a systems approach, should consider and integrate hydrological, socio-economic and environmental aspects (Jakeman and Letcher, 2003). By considering only hydrological aspects, one would ignore important socio-economic processes, which determine water demand and hence constitute actual pressure on the physical system (UNESCO, 2005; Hoekstra, 1998; Meigh, 1995).

The conceptual framework for **H**ydrological **A**ssessment of up-**S**caling **RHM** (HASR) was developed to incorporate hydrological and agricultural production systems with the aim of maximizing land and water productivity while minimizing negative hydrological and ecological impacts. The framework translates information on different adoption levels of RHM systems into simple hydrological indicators, which can be easily understood by multi-sectoral policy makers and the general public. Some examples of such hydrological indicators are the relative reduction of runoff and river flows and/or irrigation water demands due to adoption of RHM systems. The conceptual framework also explicitly brings to the fore some of the uncertainties and risks involved in making future predictions using inadequate and sometimes inaccurate data. It focuses on the impact of socio-economic development-led land use changes on water resources management at a river basin. The assumption here is that in water-scarce river basins, more water on the farm will lead to increased agricultural productivity and improved livelihoods. The framework presents different scenarios that need to be considered in the assessment of hydrological impacts of land use changes in a river basin. These scenarios consider various combinations of adoption rates of RHM and river flows regimes; high, average and low flows.

The conceptual framework is expected to inspire stakeholders to take necessary actions to address anticipated hydrological impacts and ensuing challenging water resources and livelihood issues. It will assist the stakeholders in addressing the complex process of determining the impacts of increased water retention upstream due to land use changes on downstream water users. The conceptual framework will give clarity to seemingly tenacious problems (Hajkowicz *et al.*, 2003; Giraud *et al.*, 2002; Meigh, 1995), not only in *Ewaso Ng'iro* river basin, but other water-scarce basins that are bound to experience similar problems. However, the framework would not provide simple answers to complex questions, but guide the process of formulating viable options. It is a tool to aid thinking and assist in decision-making for those responsible for developing and implementing policies. The conceptual framework would contribute to policy and institutional reforms that promote participatory approaches to integrated water resources management (IWRM), which according to Merrey *et al.* (2004) are the foundations of effective river basin level institutions.

3.3.1 Agricultural Production Systems

Understanding of farmers influence on the hydrological regime cannot be achieved without integrating their socio-economic activities and agricultural production systems that influence their decisions and actions. Intensification of rainfed agriculture is driven by the need to improve agricultural production and livelihoods. Improved agricultural productivity is measured in terms of biomass and crop yields, while livelihood is reflected in increased incomes. Besides increased yields, RHM is also aimed at stabilizing variations in crop yields and ensuring food security. However, increased production may reduce market prices and hence lower incomes, which may then either lead to a decline in adoption rate of RHM or to crop diversification. Investment in high yielding crop varieties and soil fertility improvement may also lead to increased crop yields.

The RHM production systems to be considered by HASR are soil storage systems (in-situ water conservation, micro-catchment (overland flows) and macro-catchment (diversion of ephemeral stream into cropland—spate irrigation)) and runoff storage systems (small farm ponds, medium and large storage systems such as earthdams/water pans). It is against this background that the conceptual framework for addressing the impacts of up-scaling RHM in a river basin was developed. Though Merrey *et al.* (2004), Hajkowicz *et al.* (2003) and Vincent (2003) proposed a paradigm shift from focusing on water to people who derive their livelihoods from it, the proposed conceptual framework argues for a middle ground where both water and people are the focus of IWRM strategies.

Understanding the people-water-ecosystems nexus is a prerequisite for developing sustainable IWRM strategies. This will ensure that the concerns of other people relying on the same resources and the environment are integrated. This will be ideal to reduce conflicts among winners (upstream users) and losers (downstream users) in a water resources management system. While the upstream users may justify their actions of retaining more water for productive use, the downstream users and natural ecosystems too have a right to the same water and their needs are as important. IWRM is about addressing these diverse and conflicting needs. The challenge is to support technologies that improve livelihoods of upstream farmers without compromising livelihoods of downstream users while minimizing negative hydrological and environmental impacts.

3.3.2 Spatial Mapping of RHM Systems

The location and distribution of various RHM systems in the river basin can be identified using spatial mapping based on biophysical characteristics. The spatial mapping criteria can be based on a number of bio-physical and socio-economic parameters such as soil types, soil water holding capacity, soil depth, geology, topography, local climate, land use, etc. However, to reduce the number of combinations and complexity, soil characteristics (infiltration rates) and topography (land slopes) were used as the main factors that determine the type of RHM system and hence the amount of runoff retained on agricultural lands. Land use and vegetation/crop cover are taken as the management factors, in our case, for agricultural and non-agricultural lands. Agricultural land is further categorized into traditional (no RHM systems) and improved (with RHM systems). Three sub-categories of soil characteristics and topography were used giving a total of nine combinations (Table 3.1).

Table 3.1. Spatial mapping units based on soil infiltration rates and land slopes

Soil characteristics (infiltration rates)	Topography (land slope)		
	T_L (low slopes)	T_M (medium slopes)	T_H (high slopes)
S_L (low)	S_L/T_L	S_L/T_M	S_L/T_H
S_M (medium)	S_M/T_L	S_M/T_M	S_M/T_H
S_H (high)	S_H/T_L	S_H/T_M	S_H/T_H

The spatial mapping criteria can be used to sub-divide the catchment and/or river basin based on soil infiltration rate (S) and land slope (T) and hence assign different RHM systems to a mapping unit as shown in Table 3.2. Mapping unit (i.e. hydrological response unit) is defined by the pixel sizes, which vary for different RHM systems. The three RHM systems are in-situ RHM (e.g. conservation tillage, bunds and micro-basins), small on-farm storage RHM systems (30-100 m^3 farm ponds for micro-irrigation) and medium to large storage RHM systems (earthdams and water pans mainly for irrigation and livestock water supply).

Table 3.2. Spatial mapping of different RHM systems and hydrological response unit

RHM system	Suitable sites	Pixel size	Pixel size selection criteria
In-situ RHM systems	S_H/T_L, S_H/T_M, S_H/T_H, S_M/T_H	30m x 30m	• *Landsat* image pixel size • Heterogeneity of land use
Small storage RHM systems	S_L/T_M, S_M/T_M, S_L/T_H	60m x 60m	• Minimum catchment area • Multiple of *Landsat* size
Medium-large storage RHM systems	S_L/T_L, S_L/T_H, S_M/T_L	420m x 420m	• Minimum catchment area • Multiple of *Landsat* size

For example, if $S = S_H$ (high) and $T = T_L$ (low), then in-situ RHM system is viable. However, such a spatial mapping criterion is simplistic and the decision on which RHM technology to adopt would depend on farmer's preference. Therefore, biophysical characteristics of RHM systems would be subjected to socio-economic constraints to delineate suitable land for agricultural production. The pixel sizes of each RHM system are based on the multiples of *Landsat* images pixel size (30m x 30m), catchment area and heterogeneity land use pattern and farm sizes.

3.3.3 Spatial Hydrological Scale

To understand the hydrological processes and impacts of land use changes in a river basin, one needs to analyse a river basin at different spatial hydrological scales, which forms the basis of hydrological modelling. However, hydrological data at these spatial scales are in most cases inadequate or inaccurate. Hydrological monitoring is also expensive and time consuming, hence reliance on hydrological models may be necessary. Nevertheless, hydrological models are only as good as the data used (Merrey *et al.*, 2004) and hence the main task in hydrological modelling is data acquisition, verification, analysis and validation. The different spatial scales for hydrological impact assesment are field/farm (0-5 ha), medium catchment (100-200 ha), sub-basin (20-500 km^2) and basin (>500 km^2). However, there is an increasing degree of uncertainty and complexity from field scale to river basin posing a challenge of extrapolating or interpolating results from one scale to another. Thus the need for increasing assessment from field (and runoff plot) to river basin scale to capture actual hydrological processes and agricultural production systems. Hydrological monitoring is required at field and medium catchment scales while hydro-meteorological records, where available, can provide data for sub-basin and river basin scales.

The fundamental problem related to the spatial scale issues in hydrological modelling is that hydrological systems consist of the following:

- Spatial and temporal variations in climatic condition in terms of spatial scales (varying from field to watershed, or river basin) and time scales (varying from minutes to days, months, seasons and years);
- Spatial variability of soils whose hydrological responses depend on catchment topography, geology, macro-and micro-climatic conditions which are further complicated by human activities;
- Varying vegetation with different evaporative rates, which display seasonal trends and are modified by land use changes; and
- Varying topographic features, including interlinked sub-catchments with their slope, aspect and position, which however, remain invariant over time.

This results in non-linear hydrological responses with different physical laws, which emerge and dominate at different space and time scales, of all which hydrological models try to encapsulate, often by simple and lumped calibration procedures (Schulze, 2002). Therefore, the most critical question in hydrological modelling is how best one can integrate and up-scale knowledge of micro-scale hydrological processes. Measurements at point, plot or field scales form a causal chain in order to facilitate hydrological modelling at catchment and river basin scales. More often than not, different hydrological scales would require different models (Kite *et al.*, 2001; Droogers and Kite, 2001; Kite and Droogers, 2000) due to diverse parametric variability and hydrological conditions and processes.

3.3.4 Quantification of Hydrological Impacts

The primary objective of RHM is to improve crop yields (ΔY) by increasing the transpiration component (ΔT) and reducing soil moisture stress, among other agronomic practices. Fig. 3.1 shows the three points of interventions for improving water productivity in rainfed agriculture in semi-arid environment as follows:

A: Maximizing plant water availability (maximize infiltration of rainfall, minimize unproductive water losses (evaporation from interception, soil and open water), increase soil water holding capacity and maximize root depth);

B: Maximizing plant water uptake capacity (timeliness of operations, crop management and soil fertility   management); and

C: Dry spell mitigation using supplementary irrigation (runoff storage and management).

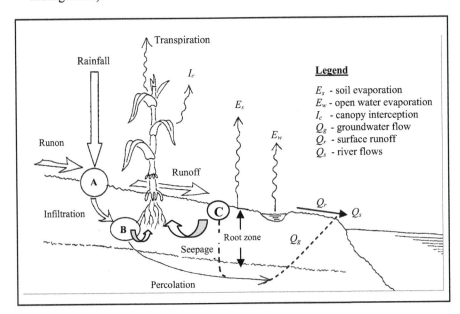

Fig. 3.1. Rainfall partitioning and intervention points (A, B & C) through RHM

The water balance analysis at these three points (A, B & C) is the basis of understanding the role of RHM in improving water productivity and food production in rainfall deficit areas. RHM may lead to: change in evaporation, $\Delta E =$ f(water surface area, interception, soil moisture); change in surface runoff, $\Delta Q_r =$ f(water retention, infiltration); change in base flow, $\Delta Q_g =$ f(deep percolation, subsurface flow, seepage); and change in soil moisture storage, $\Delta S =$ f(soil water holding capacity, rooting depth, depth of groundwater table). The hydrological and ecological impacts are reflected in change in river flow, $\Delta Q_s =$ f($\Delta Q_r + \Delta Q_g$), while socio-economic impacts are reflected in change in crop yields, ΔY. Thus quantification of hydrological impacts of up-scaling RHM is based on estimation of the components outlined in Fig. 3.1.

Rainfall-runoff relationship

Analysis of rainfall-runoff relationships in a catchment forms the basis of hydrological modelling. The relationship determines how much of the net precipitation (after subtraction of interception) is partitioned into runoff (i.e. overland flow from a catchment), some of which eventually becomes river flows, while the remainder either infiltrates into the soil on its way to the stream or is captured and stored on-farm for agricultural and domestic use. Rainfall-runoff relationships at each spatial scale indicate the amount of runoff generated by various land uses and farming systems. It shows the percentage of generated runoff that is retained on agricultural lands due to RHM—reduced catchment runoff yields—and consequently the reduction in river flow and water availability downstream. Fig. 3.2 shows linearized rainfall-runoff relationship and the effect of RHM on runoff yields at field scale. This is a typical rainfall-runoff relationship in many semi-arid areas (Ngigi *et al.*, 2005; Wairagu, 2000; Liniger *et al.*, 2000). The intercept on the horizontal axis (6.1/0.46=13 mm day^{-1}) is the amount of minimum rainfall that can generate runoff. The angular coefficient (46%) is the percentage of net rainfall that becomes runoff. The complement of the angular coefficient (54%) is the infiltration (Savenije, 1997 and 2004). For modelling purposes, the angular coefficient (runoff coefficient) would vary for different soil types and land slopes.

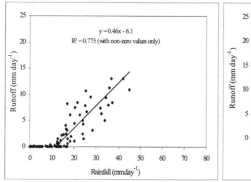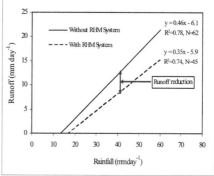

Fig. 3.2. Rainfall-runoff relationship and effect of RHM on runoff yield at farm scale

The relationship in Fig. 3.2 was developed from observed runoff data from farms with and without conservation tillage systems over four rainy seasons (2002-2003). The rainfall-runoff relationships show the amount of runoff generated and the proportion retained due to in-situ RHM systems (conservation tillage) and hence the reduction of runoff at field scale. Different RHM systems influence catchment hydrological processes at different spatial scales. In case of storage RHM systems, the amount of water stored result in a reduction in catchment yield and river flows. On the other hand, water stored for irrigation during dry seasons could reduce river water abstraction which means additional river flow, equivalent to the amount of water that would have been abstracted from the river, if runoff had not been stored.

3.3.5 Hydrological Modelling

A hydrological model should capture both negative and positive hydrological impacts related to adoption of RHM upstream. The effects of different levels of adoption will be simulated in terms of incremental water "retention" and/or "release" during rainy and dry seasons respectively. The hydrological model considers water retention/storage during rainy seasons and water "release" due to reduced abstraction, during dry season. The water "release" during dry seasons is very important because this is the time, when otherwise direct water abstraction for irrigation would drastically reduce river flows. Cases of water not reaching *Archer's Post* are increasing, and the situation is bound to get worse if adequate measures are not taken to balance upstream and downstream water needs at both spatial and temporal dimensions. According to Sivapalan *et al.* (2003), wise stewardship of water and environment requires a variety of predictive tools that can generate predictions of hydrologic responses over a range of space/time scales and climates, to underpin sustainable management of river basins, integrating economic, social and environmental perspectives.

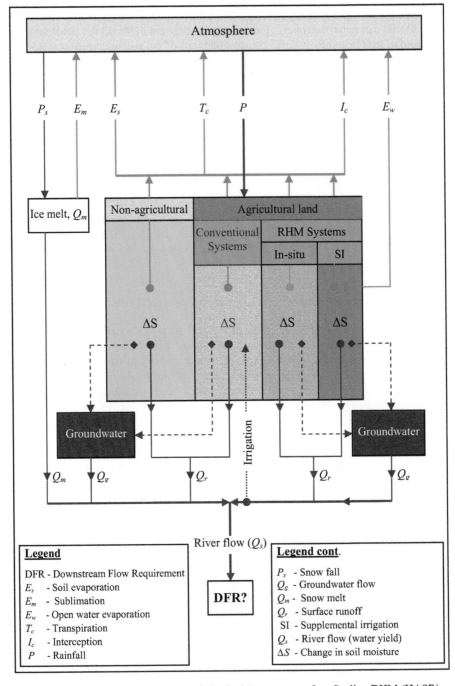

Fig. 3.3. Conceptual framework for **H**ydrological **A**ssessment of up-**S**caling **R**HM (HASR)

Fig. 3.3 presents the conceptual framework, HASR, which is an explorative model, meant to explore the hydrological implications of up-scaling RHM in a river basin. HASR analyses the amount of runoff retained by farmers upstream and hence the reduction in river flows downstream and gives emphasis on the surface runoff component that is significantly affected by up-scaling of RHM. The conceptual framework also integrates production systems on agricultural and non-agricultural lands. HASR presents a simplified conceptualization of hydrological interactions between water retention through RHM during the rainy seasons and irrigation water abstraction during the dry seasons. However, in reality the hydrological processes may not be that simple because different hydrological processes occur at different temporal and spatial scales. This notwithstanding, hydrological processes at farm/field scales influence and determine, though not directly, what happens at the larger catchment or river basin scales. In our case, hydrological monitoring at different spatial scales in 2002-2004 showed that runoff coefficient varied from 46% at field scale to 12% at river basin scale.

The amount of runoff retained by RHM system is computed using the rainfall-runoff relationships and adjusted to cater for the actual percentage of potential runoff that would be retained by each type of RHM systems (see Fig. 3.2). The decisions on the limit of up-scaling RHM are based on the amount of water available for downstream users and downstream flow requirements (DFR). If the river flow is below the minimum downstream flow requirements, decisions will be made to reduce the amount of runoff retained by farmers and thus the limit of up-scaling RHM will have been reached under that type of production system.

From spatial mapping of RHM systems (section 3.3.2) in the river basin, the effects of each system can be quantified from the runoff reduction ratio—proportion of runoff retained on agricultural lands. Then the runoff yields reduction ratio by a RHM system ($\Delta Q_{r(S)}/Q_{r(S)}$) is calculated from Eq. (3.1) and the total runoff reduction ratio in a catchment or river basin ($\Delta Q_r/Q_r$) is computed from Eq. (3.2). Up-scaling of RHM is reflected by an increase in area under RHM (ΔA_S) and the intensification of RHM leading to more runoff retention per unit area ($\Delta q_r/q_r$). Runoff reduction is computed per unit area based on the spatial scale (i.e. 1 ha for in-situ and small storage systems and 1 km^2 for medium to large storage systems).

$$\frac{\Delta Q_{r(S)}}{Q_{r(S)}} = -\left(\frac{A_s}{A_R}\right)\frac{\Delta q_r}{q_r} \qquad\qquad (3.1)$$

$$\frac{\Delta Q_r}{Q_r} = \sum\left(\frac{\Delta Q_{r(SS)}}{Q_{r(SS)}} + \frac{\Delta Q_{r(IS)}}{Q_{r(IS)}}\right) \qquad\qquad (3.2)$$

where; A_S = area under a RHM system (ha), A_R = total area of the catchment or river basin (ha), $\Delta q_r/q_r$ = runoff retention per unit area, $\Delta Q_{r(S)}/Q_{r(S)}$ = runoff reduction ratio by a RHM system (%), $\Delta Q_{r(SS)}/Q_{r(SS)}$ = runoff yield reduction ratio by storage RHM systems (%), $\Delta Q_{r(IS)}/Q_{r(IS)}$ = runoff yield reduction ratio by in-situ RHM systems (%), and $\Delta Q_r/Q_r$ = total catchment/river basin runoff reduction (%).

Irrigation water abstractions have been identified as one of the main contributing factors to reduced river flows, especially during the dry periods when many farmers along the streams abstract water illegally and uncontrollably without due regards to downstream water users. The irrigation-RHM interface presents a positive effect on irrigation water supply, in terms of reduced dry season river abstractions. This is based on the "released" water that would have been drawn from river flows if runoff was not harvested and stored during the rainy seasons. Medium and large RHM systems may be viable options for reducing dry season irrigation water abstraction. Thus RHM systems may reduce water scarcity related conflicts among upstream and downstream users.

3.4 Results and Discussion

3.4.1 The Case of Naro Moru River Sub-basin

Background information on *Naro Moru* sub-basin is presented in section 2.2.2. The river has six river gauging stations from the top of Mt. Kenya to the point where it joins *Ewaso Ng'iro* river (Fig. 3.4). *Naro Moru* sub-basin is divided into sections (reaches) according to elevation and ecological belt i.e. moorland (3500-5200m), forest (2300-3500m), foot zone (2000-2300m) and savannah (1800-2000m). Much of the river flow is concentrated within the two rainy seasons. River discharges during the dry months consist mainly of base flow, which is derived from groundwater sources in the lower moorland and upper forest zones. The semi-arid savannah only yields runoff during the rainy season.

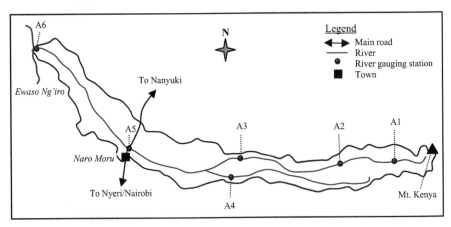

Fig. 3.4. The six river gauging stations (A1-A6) of *Naro Moru* sub-basin

Water abstraction assessment revealed that about 62% of the dry season flow and 43% of the wet season flow is abstracted from *Naro Moru* river before its confluence with *Ewaso Ng'iro* river (NRM, 2003). This shows heavy utilization of river flows through abstractions to support human, livestock and irrigation. Though the river is perennial, over-abstraction, of which more than 70% is illegal (Aeschbacher *et al.*, 2005; Gichuki *et al.*, 1997 and Gikonyo, 1997), leads to drying up of the lower reach during the driest months of February and March, and under extreme conditions from July-September. River flow in the dry seasons is only enough for domestic and livestock water needs and for micro-irrigation. Irrigation water demand can only be met if RHM—construction of storage reservoirs—is considered. Over the years, it is becoming evident that continued land-use changes, especially on the foot zone and savannah (mainly due to agricultural development), is affecting river flows. The average river flows on the lower river reaches (foot and savannah zones) have gradually been decreasing, as shown in Fig. 3.5, while the upper reach (forest zone) indicates no significant decline.

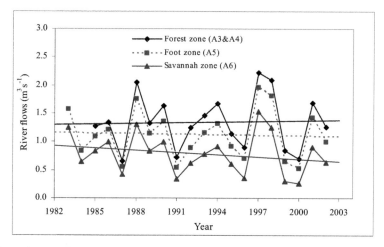

Fig. 3.5. Observed flows and trends along three reaches of *Naro Moru* river

The management of diminishing water supply poses a major challenge due to related hydrological, environmental and social implications. This calls for proper water management to ensure that this resource is used in a sustainable way. In the past, emphasis has been on supply of river water to meet demand but there is an urgent need to devise viable options to manage the increasing demand. Some of the options include improving water use efficiency, soil moisture conservation in rainfed agriculture, restricting water use during critical dry periods and storage for use during the dry seasons. However, sustainable solutions to addressing conflicts over water rely on formulation of adaptive policies and strategies.

3.4.2 Anticipated Scenarios and Hydrological Impacts

Fig. 3.6a shows a decreasing trend of *Ewaso Ng'iro* river flows at *Archer's Post,* which can be attributed to land use changes and upstream water abstraction mainly for irrigation, since there has been no significant reduction in rainfall. This reflects what is happening at the sub-basins upstream, for example the case of *Naro Moru* presented in Fig. 3.5. Therefore, the anticipated scenarios shown in Fig. 3.7 are based on two hypotheses: (i) adoption of RHM will increase progressively due to its tangible benefits to farmers; and (ii) increased retention of runoff upstream will reduce river flow.

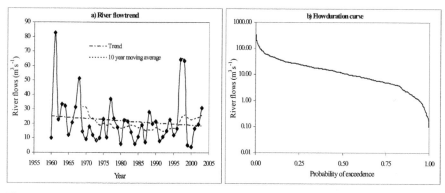

Fig. 3.6. Flow trend and duration curve of *Ewaso Ng'iro* river at *Archer's Post*

The scenarios of up-scaling RHM are based on three river flow regimes; high (upper quartile), average (median) and low (lower quartile) flows, i.e. at 25% (Q_{25} = 25.16 m^{3}s^{-1}), 50% (Q_{50} = 11.53 m^{3}s^{-1}) and 75% (Q_{75} = 4.97 m^{3}s^{-1}) probability of exceedence respectively. These probabilities are based on observed historical river flow data of *Ewaso Ng'iro* river at *Archers Post* (see Fig. 3.6b) and anticipated adoption rates of RHM systems. The y-axis on the right represents the observed long term river flow while that on the left represents anticipated rate of adoption of RHM systems. The adoption rate, currently estimated at 15%, is based on increased awareness of RHM and related impacts on agricultural productivity and household livelihoods. Adoption of RHM can also be attributed to diminishing river flows, which has prompted commercial farmers to construct runoff storage reservoirs for irrigating high value horticultural crops (Ngigi, 2003a).

The intersection point of river flows and the desired DFR indicates the limit of up-scaling RHM. Even at the present low RHM adoption rate, there are cases of water not reaching the basin outlet during extremely low rainfall years. The progressively decreasing river flows (see Fig. 3.6a) may be attributed to land use changes, and increasing adoption of RHM, which is recommendable until the up-scaling limit is exceeded—river flows fall below the desired DFR. In general, the river flow patterns show short peak flow and long low flow periods (Gichuki, 2002; Liniger *et al.*, 2005). This means that the river flow during dry season is some times below the minimum DFR, which for our case can be conservatively estimated as Q_{95} (i.e. 0.95m^{3}s^{-1}). Reduced river flows could lead to negative hydrological, socio-economic and environmental impacts for downstream water users. High river flows during the rainy seasons are important for recharging groundwater and maintenance of natural ecosystems. These anticipated river flows reduction present a big challenge and hence the need to formulate sustainable solutions.

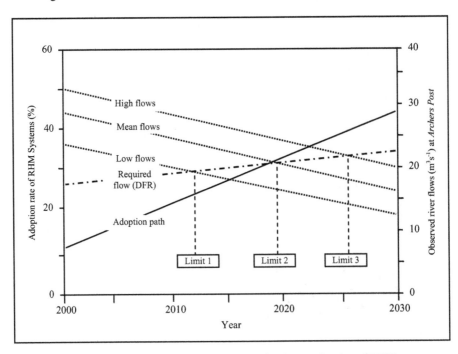

Fig. 3.7. Simplified presentation of anticipated scenarios due to adoption of RHM systems

The long-term hydrological impacts can be assessed by simulating the HASR management scenarios using established river basin hydrological models such as the soil and water assessment tool (SWAT). SWAT is a physically-based continuous-event hydrological model for predicting the impacts of land management practices on water, sediment and agricultural chemicals in large complex watersheds with varying soils, land use and management conditions over long periods of time (Nietsch *et al.*, 2001). The SWAT model can simulate different land use scenarios related to up-scaling RHM as conceptualized by HASR in Fig. 3.3 and hence hydrological impacts on downstream water resources management. Thus HASR can be integrated into complex hydrological models to enhance long-term assessment and policy formulation. However, caution should be taken as a more sophisticated model may not alone solve the impediments of data quantity and quality, uncertainty and scaling issues (Sivapalan *et al.*, 2003). Nevertheless, with substantial data, it is imperative to apply hydrological modelling to enhance formulation of sustainable IWRM policies.

3.4.3 Formulation of IWRM Policies

Policy and decision makers are faced with a dilemma due to inadequate information and simple methodologies for assessing the consequences of socio-economic development, which bring about land use changes and hydrological impacts on water resources management. In the past, water has been perceived merely as a "free" resource, to be exploited in order to support socio-economic development. Impacts of human activities on water resources management are manifested in limited water supply and increasing tension between intensive water use and the functioning of natural ecosystems. Moreover, uncertainty and risks associated with forecasting future scenarios and trends on the utilization of natural resources affect policy formulation process.

HASR can address this by highlighting possible consequences of stakeholders' actions or inactions towards land and water development. The framework provides a tool for assessing hydrological impacts based on various production systems and guide formulation of IWRM policies at a river basin scale. It forms the basis of formulating policies and strategies by highlighting key issues and the process of understanding them. Therefore, the framework can enhance decision support system (DSS) and stakeholders' dialogue. DSS would be based on optimization of production systems (adoption of RHM systems) and hydrological impacts assessment related to land use changes. DSS provides the stakeholders with options and related hydrological impacts thus feeding into stakeholders' dialogue. The dialogue focuses on trade-offs among conflicting water users' interests. The conceptual framework integrates the needs and aspirations of different stakeholders (and their production systems) and the needs of natural ecosystems i.e. hydrological, socio-economic and environmental aspects. It aims to maximize land and water productivity of upstream and downstream users, minimize conflicts among water uses, and maintain minimum flows for conservation of natural ecosystems and downstream users beyond the river basin boundaries. Analysis of the "best" or "acceptable" land and water management options will consider the results of optimization and evaluation of trade-offs among different stakeholders and socio-economic sustainability of production systems (Mainuddin *et al.*, 2003).

Sustainable water resources management strategies aim to improve livelihoods of downstream and upstream inhabitants and conservation of natural ecosystems. In tackling complex problems of IWRM, the main challenge is how to integrate pertinent policy, legal and institutional issues. Problems with livelihoods, food security, unequal distribution of resources and income, depletion of natural resources and ecosystems conservation are some of the issues stakeholders have to address in formulating responsive land use policies, legislations and enforcement mechanism to realize IWRM under water scarce situations. The era of formulating sectoral policies and legislation is coming to an end as water resources become scarcer, hence affecting other sectors of development. Hence, formulation of sustainable water resources management policies needs an integrated and multi-sectoral approach.

Hydrological modelling can be used to assess impacts of upstream land use changes related to increased adoption of RHM for improving rainfed agriculture. The results can show how socio-economic activities of upstream land users can affect their downstream counterparts, and what needs to be done to address such impacts. The main stakeholders are the downstream and upstream land and water users, who strive to make maximum benefits from a resource they have always considered as free and a common good. When the resource is adequate, there is no conflict and hence status quo remains. However, more upstream water abstraction leads to water scarcity downstream. Conflicts among different water users may lead to social disruptions, which would affect socio-economic development.

Different water users perceive the problem differently. The downstream users see upstream irrigators as the problem. The irrigators on the other hand argue that it is the only way they can produce food in this semi-arid environment. The problem is bound to be more complicated if improved rainfed agriculture (e.g. RHM systems) could retain more water upstream. HASR gives more emphasis irrigation-RHM interface that could enhance water balance and reduce conflicts.

3.5 Conclusions

Land use changes, especially intensification of rainfed agriculture, are unavoidable due to increased food demand and declining agricultural productivity. Such changes are bound to have positive socio-economic impacts geared towards improving livelihoods, but could lead to negative impacts downstream. This would affect downstream livelihoods and natural ecosystems that depend on sustained river flows. HASR analyzes the anticipated scenarios and socio-economic, hydrological and environmental impacts on upstream and downstream reaches of *Ewaso Ng'iro* river basin. The socio-economic impacts are based on improved agricultural production through adoption of RHM systems, which retain and store more water for crop production. However, this may reduce runoff and hence river flows, which may lead to water scarcity downstream. HASR can be used to asses the impacts on up-scaling RHM, and hence form the basis hydrological modelling and formulation of IWRM strategies.

Up-scaling of RHM can be attributed to increased agricultural production and stabilized crop yields, and hence improved income and livelihoods. However, there is need for preparedness to address anticipated impacts and resulting water crisis. This forecast will assist stakeholders to formulate sustainable policies to avert the looming water crisis. Moreover, there is need for more detailed predictions of the possible scenarios. Application of advanced hydrological models to simulate anticipated scenarios would be one method of achieving this. Nevertheless, the preliminary results guide the process of formulating IWRM strategies to address anticipated land use changes and related impacts. Thus HASR will play a role in answering the following question "what is the limit of up-scaling RHM in a river basin scale?" Ngigi (2003b). The answers form the basis of sustainable IWRM strategies, especially for water-scarce basins

A sustainable IWRM strategy should balance the diverse interests of stakeholders. It is envisaged that HASR would enhance understanding of various hydrological and socio-economic processes and hence support formulation of sustainable policies, legislation and institutions that focuses on the needs and socio-economic activities of water users and natural ecosystems. The policy formulation process requires an understanding of trade-off between land and water systems as well as potential impacts on other sectoral policies. This would identify sustainable options for water resources management.

The conceptual framework defines how land use changes and hydrological impacts can be integrated in a decision support system. It is an explorative tool aimed at strategic land use issues: how to satisfy the conflicting needs and objectives on economic, food security, ecological and social dimensions of land use. Its primary aim is to support and stimulate open discussion about future possibilities and limitations. HASR is simple enough, for the stakeholders to understand the hydrological impacts of land use changes, and comprehensive enough to predict possible future scenarios and trends under different land use and water management systems. Besides contributing to the on-going restructuring of water resources management in Kenya, the paper will contribute to international policy dialogue and programs such as Hydrology for the Environment, Life and Policy (HELP) (UNESCO, 2005), Consultative Group on CGIAR Challenge Program on Water for Food (IWMI, 2002), among others.

Chapter 4

4.0 Agro-hydrological Assessment of On-farm Storage RHM Systems[7]

4.1 Overview

Semi-arid agro-ecosystems are characterized by erratic rainfall and high evaporation rates leading to unreliable agricultural production. Total seasonal rainfall may be enough to sustain crop production, but its distribution and occurrence of intra- and off-season dry spells affect crop production. RHM, especially on-farm storage ponds for supplemental irrigation offers an opportunity to mitigate the recurrent dry spells. Farm ponds are small runoff storage structures of capacities ranging from 30-100 m³ used mainly for supplemental irrigation of kitchen gardens, and sometimes for domestic and livestock water supply. The main objective was to evaluate the hydrological and economic performance of farm ponds with the view of assessing their contributions to water and food security in semi-arid agro-systems of Kenya. Agro-hydrological evaluation of on-farm runoff storage systems entailed field survey, monitoring of water losses, analysis of rainy seasons and dry spell occurrence, soil moisture and water balance, estimation of supplemental irrigation requirement (SIR) and farm-level benefit-cost analysis of cabbage production using low-head drip irrigation system. Significant water losses through seepage and evaporation, which accounted on average for 30-50% of the stored runoff, is one of the factors that affect the adoption and up-scaling of on-farm water storage systems.

Frequency analysis of rainfall revealed that there is 80% probability of occurrence of dry spells exceeding 10 and 12 days during the long rains and short rains respectively. The occurrence of off-season (after rainfall cessation) dry spells was more pronounced than intra-seasonal (within the rainy season) dry spells. The length of intra-seasonal (10-15 days) was less than off-season dry spells (20-30 days). The occurrence of off-season dry spells coincides with the critical crop growth stage, in particular flowering and yield formation stages. A 50 m³ farm pond with a drip system irrigation system was found adequate to meet SIR for a kitchen garden of 300-600 m² planted with a 90 days growing period cabbages. The benefit-cost analysis showed that farm ponds are feasible solutions to persistent crop failures in semi-arid areas which dominant most countries in SSA.

Improvement of rainfed agriculture, which constitutes more than 80% of Kenyan agricultural production, has been identified as one of the key solutions to persistent food insecurity in the ASALs of SSA, and Kenya in particular. The ASAL are characterized by erratic rainfall with high annual variability, annual

---

[7] *Based on*: Ngigi, S.N., H.H.G. Savenije, J.N. Thome, J. Rockstrom and F.W.T. Penning de Vries. 2005. Agro-hydrological evaluation of on-farm rainwater storage systems for supplemental irrigation in *Laikipia* district, Kenya. *Agricultural Water Management, 73 (1): 21-41*

potential evaporation exceeding the rainfall amounts, high amounts of runoff due to low infiltration and recurrent soil moisture deficits limiting crop production (Ben-Asher and Berliner, 1994; Perrier, 1988; Evenari *et al.*, 1971). Under these conditions, rainfed agriculture, which is one of the main economic activities, has failed to provide minimum food requirements for the rapidly increasing population.

One of the challenges for improving livelihoods in the ASALs is upgrading rainfed agriculture which contributes 30-40% of GDP (World Bank, 1997), 90% of food supply (Savenije, 1999) and covers more than 95% of croplands in water scarce tropics of SSA (FAO, 2002). However, rainfed agriculture in SSA is characterized by low productivity due to sub-optimal performance related to management aspects rather than low physical potential (Rockström and Falkenmark, 2000 and SIWI, 2001). Although semi-arid environments of SSA are characterized by poorly distributed rainfall, crop production is not necessarily affected by absolute water scarcity, but recurrent dry spells. Nevertheless, RHM can ensure optimal crop production in an area with inadequate rainfall. RHM offers farmers an opportunity to improve agricultural production, especially bridging intra-seasonal and off-season dry spells. For instance, marginal lands with rainfall as low as 300 mm yr^{-1} can be made productive if controlled and limited water is made available by RHM techniques (Flug, 1981 and Ngigi, 2003a).

However, although RHM are promising interventions towards increasing and sustaining agricultural production in the ASAL, widespread adoption and up-scaling could have hydrological implications at catchment and river basin scales and consequences on overall river basin water resources management. In the past, a number of RHM projects have been set up with the objective of combating the effects of intra-seasonal drought. However, only a few projects have succeeded in combining technical efficiency with low cost and acceptability to local farmers (Reij *et al.*, 1988 and Critchley, 1999). This has been partially attributed to the limited technical knowledge and inappropriate technology dissemination approaches with regard to the prevailing socio-cultural and economic conditions (Ngigi, 2003a). However, impacts from such projects can be realized through hydrological evaluation and improvements of RHM technologies.

A challenge in design and construction of on-farm water storage structures such as farm ponds is how to minimize water losses (mainly due to seepage and evaporation). The amount of water lost through evaporation depends directly on the evaporation rate and the reservoir surface area. Whilst seepage losses from a storage system will vary with the soils, the type of underlying rock, the time that water is held and the volume of water stored. The acceptable seepage losses depend on intended use and water management. Generally the limited stored runoff by smallholder on-farm ponds may not permit full irrigation of most crops. Instead water is used for supplemental irrigation of rainfed crops to mitigate dry spells and/or full off-season irrigation of small-scale vegetable gardens. Supplemental irrigation is also applied to mitigate intra-seasonal dry spells which may occur during critical crop growth stages. Barron *et al.* (2003) showed that in any year, there is a 70% probability of a dry spell exceeding 10 days during flowering for maize in semi-arid *Machakos* (Kenya) and *Same* (Tanzania) districts.

In ASAL, periods of severe water stress are common and often coincide with the most sensitive stages of growth making RHM for supplemental irrigation a particularly interesting management system for rainfed agriculture. The principles of moisture balance analysis on short time steps is normally seen as very complicated, due to the dynamic and complex process involved when water flows through saturated and unsaturated soils (Rockström, 2000b). However, simple approaches have been used for the purpose of planning for RHM. In order to assess water needs to bridge or mitigate dry spells, a detailed water balance analysis is needed. The accounting should be based on a time step that actually captures the occurrence of dry spells. The time steps should be ideally on a daily basis, but for practical purposes, less than 10 days (Rockström, 2000b) depending on the length of intra-seasonal dry spells.

The main objective of the study was to evaluate the hydrological and economic performance of farm ponds with the view of assessing their contributions to addressing the persistent water and food crises in most semi-arid agro-systems of SSA and in particular Kenya. The chapter presents the results of a farm-level agro-hydrological evaluation of on-farm storage ponds, especially their performance in terms of bridging both intra- and off-season dry spells which occur during the crop growing period. It addresses dry spell occurrence and mitigation through supplemental irrigation using on-farm rainwater storage systems. Moreover, water application using drip irrigation and benefit-cost analysis of a medium kitchen (vegetable) garden planted with cabbages are assessed. The detailed study was carried out in semi-arid *Matanya* area, Central division of *Laikipia* district in upper *Ewaso Ng'iro* river basin of Kenya.

4.2 Methodology

4.2.1 Site Description

The impact of RHM on agricultural production was assessed in central parts of *Laikipia* district, which is situated within the transitional zone from wetter to drier climatic regime. The location of the study area is shown in Fig. 4.1. Rainfall ranges between 280-1100 mm yr^{-1} and the mean annual temperatures range between 16-20^{0}C (Berger, 1989). The rainfall pattern is bi-modal with the long rainy season occurring from March to May and the short rainy season from October to November (see Fig. 4.2). The area also receives continental rains, which occur between June and September. However, agricultural production is constrained by soil moisture deficit associated with unreliable and poorly distributed rainfall, which is characterized by periodic intra- and off-season dry spells. The soils are mainly dark grey clay, i.e. *Verto-luvic Phaeozem* (Liniger, 1991; Desaules, 1986) whose characteristics are presented in Table 4.1.

Table 4.1. Soil characteristics in *Matanya, Laikipia* district

Depth (cm)	Textural class (%)				AWC (vol. %) (i.e. mm/100mm)	AWC (mm)
	Clay	*Silt*	*Sand*	*Organic matter*		
0 - 20	40	22	34	4	20.0	40
20 - 45	52	20	27	1	16.2	41
45 - 75	64	15	21	0	18.0	54
75 - 100	58	14	28	0	16.9	42
100 - 130	47	23	30	0	12.4	37
130 - 160	47	18	35	0	10.7	32
160 - 180	47	18	35	0	10.7	21

Note: AWC = Available water content

Land use has been changing since the early 1960s from previous large-scale livestock production in former European farms to small-scale subsistence agricultural production systems. Most of the small-scale farmers migrated from the adjacent highly populated districts due to pressure on land. Transfer of agricultural production systems from high potential areas to these semi-arid environments has led to low productivity and environmental degradation. However, with assistance from the government and external support agencies, the farmers have been adopting various RHM systems to address water scarcity and reduce periodic crop failures. On-farm runoff storage ponds, which are mainly used for supplemental irrigation of kitchen gardens, are one of the most common RHM systems in the study area. In the context of this study, a farm pond refers to small truncated coned-shaped subsurface storage structures of 30-100 m³. Normally farm ponds would refer to storage structures of up to 1,000 m³. Desspite the significant impact of this type of RHM system, one of the major challenges is how to effectively utilize the limited amount of water harvested and stored by the small on-farm storage ponds.

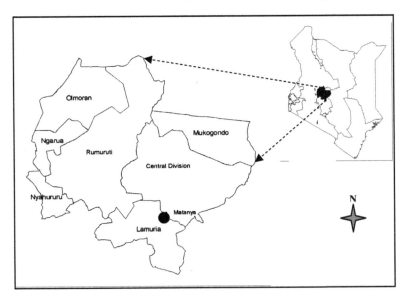

Fig. 4.1 Location of *Matanya* in *Lakipia* district of Kenya

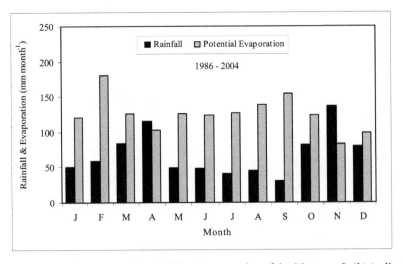

Fig. 4.2. Long term mean monthly rainfall and evaporation of the *Matanya, Laikipia* district

4.2.2 Evaluation of On-farm Storage RHM Systems

Field survey

The locations in the district where on-farm storage RHM systems have been introduced were identified from the Ministry of Agriculture records. A field survey was carried out to evaluate the performance of a number of farm ponds (water pans) in the district. The field data were collected through observations, physical measurements, interviews and existing records. Pre-determined innovative farmers also provided useful information. The field survey focused on the frequency of water harvesting and storage, amount of water stored, duration of water storage, water uses and application methods, impacts on food production and farmers' experiences and constraints. Catchment areas, storage capacities and areas under irrigation were measured. Field observations were made on general conditions of the farm ponds, crop appearances, soil characteristics and types, vegetation and water application methods. The field evaluation was done during rainy seasons and following dry seasons. Field sites for detailed agro-hydrological evaluations were identified in collaboration with the farmers. Site selections were based on rainfall distribution, soil and catchment characteristics, condition and history of the RHM systems, water management, crop grown, farmers' experiences and willingness to participate in the study.

Experimental design

In each experimental site, two farm ponds were selected; natural (unlined) and lined (with ultra-violet resistant plastic). The catchment areas were demarcated and sizes determined. Water inflows and outflows were monitored daily and results used for the water balance analysis. Runoff monitoring was accomplished by using pipe samplers and/or water level recorders—each site had one farm pond with both devices. The pipe sampler is a non-mechanical runoff measuring device which is designed on the principle of uniform fluid flow (Hai *et al.*, 2004). It is a rectangular-shaped metallic channel (242 mm wide, 156 mm high and 465 mm long) made of 2 mm thick mild steel as shown in Fig. 4.3.

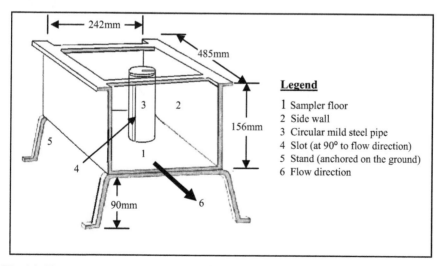

Legend

1 Sampler floor
2 Side wall
3 Circular mild steel pipe
4 Slot (at 90° to flow direction)
5 Stand (anchored on the ground)
6 Flow direction

Fig. 4.3. The components of a pipe sampler used for runoff measurement

Two 100 mm² bars are welded at the top to hold the sides perpendicular to the base. A 165 mm long mild steel pipe of 60 mm diameter with a 140 mm slot of 5 mm is welded at the centre of the channel base with the slot perpendicular to the direction of flow. The slotted pipe is fitted with a cover flap. A similar pipe is fitted at 45° from 30 mm diameter outlet of the slotted pipe through which an average of 1% (0.5-1.5%) of the total runoff is collected and measured (Hai *et al.*, 2004). Flow through the slot is influenced by surface roughness, gradient, flow velocity and patterns. Experience shows, that the slot collects slightly more runoff (i.e. >1%) at lower flow velocities than at high velocities (Hai *et al.*, 2004). A pipe sampler, while giving slightly less reliable data than a tipping bucket, is about 10 times cheaper, and thus an appropriate alternative for extensive runoff studies (Zöbisch *et al.*, 2000; Khan and Ong, 1997).

Water outflows and inflows were monitored daily (normally at 9.00am and 300pm). Water level monitoring was done to determine water outflows—irrigation water use and water losses (evaporation and/or seepage). Any other water withdrawals were also monitored by recording how much water was used daily. Rainfall was recorded by standard rain gauges, while evaporation was monitored by Class A evaporation pans installed at the experimental sites. On rainy days the water depths in the farm ponds were recorded before and after the rains.

4.2.3 Water Balance Analysis

The agro-hydrological parameters of the common truncated cone shaped farm ponds used in the water balance analysis (Eq. 4.3) are illustrated in Fig. 4.4. The dimensions of the farm pond, which include the maximum water depth, top and bottom widths and the side slope, were measured directly and used to compute the storage capacity. However, the dimensions varied from one farm pond to another.

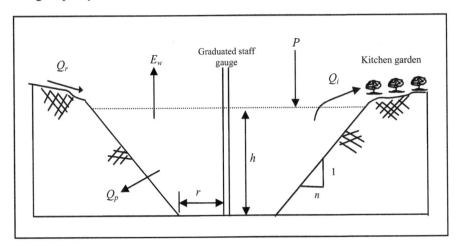

Fig. 4.4. Typical farm pond showing agro-hydrological parameters

The volume and exposed surface area of the farm ponds are expressed as functions of depth of water by applying solid geometry Eqs. (4.1 & 4.2) developed by Helweg and Sharma (1983).

$$V = \frac{1}{3}\pi h\left[3r^2 + 3nhr + n^2h^2\right] \tag{4.1}$$

$$A = \pi(r + nh)^2 \tag{4.2}$$

where; V = storage capacity of the pond (m³), A = exposed surface area (m²), h = water depth (m), n = side slope^{-1} (Fig. 4.4), and r = bottom radius of the pond (m).

Eqs. (4.1 & 4.2) and the respective farm pond dimensions were used to determine the on-farm water balance of the RHM system as shown in Eq. (4.3).

$$\frac{dV}{dt} = (Q_r + PA) - (Q_i - E_w A - Q_p) \tag{4.3}$$

where; V = storage of the reservoir (m³), Q_r = surface runoff (m³ s^{-1}), P = direct precipitation (m s^{-1}), Q_i = irrigation requirement (m³ s^{-1}), E_w = open water evaporation (m s^{-1}), A = exposed surface area (m²), Q_p = seepage losses (m³ s^{-1}), and t = time (s).

The changes in water level were recorded before and after each rainfall event. The increase in water level was attributed to runoff and direct rainfall (inflow) and seepage over the rainfall duration (outflow). Water levels, evaporation pan and rain gauge readings were recorded daily at 9 am and 3 pm. The water levels were converted to volume and exposed surface area respectively using depth-volume and depth-surface area relationships in Eqs. (4.1 & 4.2). The exposed surface area determines the evaporation losses, while seepage losses depend on wetted surface area and soil characteristics. Seepage losses were computed from the water balance analysis using Eq. (4.3), in which all the other parameters are measurable. Besides, the results were compared with those from lined (no seepage losses) and unlined farm ponds to reduce errors that could be associated with the measurable parameter. Each site had two farm ponds, one lined with plastic sheet to control seepage and the other one unlined for the purpose of verifying the water balance results.

4.2.4 Analysis of Rainy Seasons and Dry Spells

The occurrence of rainy seasons and dry spells (intra- and off-season) was carried out based on frequency analysis of 16 years (1987-2002) daily rainfall and evaporation data for *Matanya* rainfall station within the study area. The onset of rainfall was determined by combining the FAO (1978) and Berger (1989) criteria. The FAO (1978) criteria define the onset of rainy season as the time of year when precipitation equals or exceeds $0.5E_o$. Berger (1989) qualifies this further by considering a 15-day period during which if the cumulative rainfall within 5 consecutive days is more than 20 mm, and the next 10 days also receive more than 20 mm, then, then the first day of the period marks the onset of rainy season. Similarly, rainfall cessation is defined as the time when precipitation falls below $0.5E_o$. However, within the defined rainy season, intra-seasonal dry spells occur, and if crop growing period extends beyond the rainy season, off-season dry spells occur that may affect the crop at maturity stage. These dry spells affect crop production depending on their timing and magnitude with respect to crop growth stages and sensitivity to water stress. The length of off-season dry spells was determined by superimposing actual crop water requirements based on growth stages and transpiration demand over the growth period starting at the onset date.

The analysis of occurrence of rainy season and dry spells was carried out for each year to determine the onset, cessation and length of dry spells. Frequency analysis of 16 years record yielded the probability of occurrence and exceedence of rainy seasons and length of dry spells respectively for both the long rains (March-

May) and short rains (October-December). The effect of staggering the planting dates within the rainfall onset window on occurrence of dry spells was also analyzed by considering different occurrence probabilities i.e., early (start of onset window (25%), optimal onset (50%) and late (end of onset window (75%) planting. The probability of occurrence of a dry spell exceeding 10 and 15 days was determined for different planting dates. The adequacy of seasonal rainfall over the entire growth period was analyzed based on crop water requirements and soil moisture balance. When intra- and/or off-season dry spells occur, supplemental irrigation is applied to bridge the soil moisture deficit. In this case only adequate water for the most sensitive (critical) growth stage, in which water shortage would drastically affect yields, is applied. The SIR depends on the crop water requirement and soil moisture deficit. The daily SIR was superimposed on rainfall and crop water requirement to show the timing and amount of SIR.

4.2.5 Determination of SIR

Crop water requirement

The potential evaporation (E_o) was estimated from evaporation data from a *Class A Pan* installed next to the farm ponds. A pan coefficient value of 0.8 was used to convert the pan evaporation (E_{pan}) to potential evaporation (E_o). The crop water requirement (E_c) for cabbage with 90-day growing period was determined according to Doorenbos and Pruitt (1977) as follows:

$$E_c = K_c * E_o \qquad\qquad\qquad (4.4)$$

$$E_o = K_p * E_{pan} \qquad\qquad\qquad (4.5)$$

where; E_c = crop water requirement (mm day^{-1}), E_o = potential evaporation (mm day^{-1}), K_c = the crop factor, and K_p = the pan factor/coefficient.

SIR for cabbages

The SIR was computed by applying a water balance approach to soil moisture balance analysis, which was based on the following assumptions:

- There is no groundwater contribution and deep percolation (Q_{dp}) below root zone occurs when the soil moisture exceeds the soil field capacity.
- Planting is done at the onset of rainfall, and the initial soil moisture reserve in the planting 10-day interval is taken to be equal to half the available soil moisture in the rooting depth.
- The rooting depth (D_{rz}) for cabbages is taken as 20, 40 and 60 cm for initial stage (15 days), crop development (20 days) and maturity stages (from mid-season) (55 days) respectively.
- The crop co-efficient (K_c) for cabbage of 0.45, 0.75, 1.03 and 0.95 for initial (15 days), development (20 days), mid-season (45 days) and late season (10 days) stages are used.
- Drip irrigation with irrigation efficiency of 90% was adopted.

The soil moisture balance was carried out on a 10-day time step with the start of the first time step corresponding to the planting duration (rainfall onset). The 10-day time interval was determined from frequency analysis at 80% probability of exceedence of the length of dry spells. For each 10-day time step the soil moisture balance of the unsaturated zone was carried out using the following water balance equation:

$$\frac{S_t - S_{t-1}}{\Delta t} = P_e - T_c - E_s \tag{4.6}$$

where; S_t = soil moisture storage per unit surface area at time t (mm), S_{t-1} = soil moisture storage per unit surface area at time $t-1$ (mm), Δt = time interval (10-day period), P_e = effective rainfall (mm 10-day^{-1}), E_s = soil evaporation (mm 10-day^{-1}), and T_c = crop transpiration (mm 10-day^{-1}).

Effective rainfall (P_e) is the proportion of rainfall (P) that is either directly, as soil moisture stored in the root zone, or indirectly, as surface runoff that is stored in reservoirs and applied through irrigation, available for plants use. Therefore, P_e incorporates surface runoff, interception and soil moisture storage (infiltration) within the root zone. Part of deep percolation (Q_{dp}) that can be utilized by the plants through capillary rise is also included. However, under extreme high rainfall, not all deep percolation can be accounted for by effective rainfall. P_e was estimated for our conditions using empirical equations (Dastane, 1974; FAO and WMO, *Undated*), which were adjusted, by dividing the y-intercept and P limits by 3 to take care of 10-day interval instead of monthly rainfall values as follows:

$$P_e = 0.6P - 3.3 \quad \text{for } P < 23.3 \text{ mm 10-day}^{-1} \tag{4.7}$$

$$P_e = 0.8P - 8.0 \quad \text{for } P > 23.3 \text{ mm 10-day}^{-1} \tag{4.8}$$

where; P is rainfall (mm 10-day^{-1} and P_e = effective rainfall (mm 10-day^{-1}).

If there is soil moisture deficit (S_d) to meet the crop water requirement, i.e. $S_d > 0$, then soil moisture is withdrawn from the root zone storage accrued during the previous time step (S_{t-1}). In case of excess rainfall, there is a negative deficit (i.e. – S_d) and hence soil moisture is added to storage. If $S_d \Delta t \le S_{t-1}$ then there is enough soil moisture available to meet the deficit and hence $Q_i = 0$; else, moisture stored in soil cannot meet the crop water requirement and supplemental irrigation is required. The seasonal supplemental irrigation requirement (Q_i) is calculated as follows:

$$Q_i = (T_c + E_s + Q_{dp}) - (P_e + \frac{S_{t-1}}{\Delta t}) \tag{4.9}$$

where; Q_i = supplemental irrigation requirement (mm season^{-1}), S_t = soil moisture storage per unit surface area at time, t (mm), S_{t-1} = soil moisture storage per unit surface area at time, $t-1$ (mm), Δt = time interval (season, e.g. 90 days), P_e = effective rainfall (mm season^{-1}), Q_{dp} = deep percolation (mm season^{-1}), E_s = soil evaporation (mm season^{-1}), and T_c = crop transpiration (mm season^{-1}).

In case of surplus soil moisture (S_s) i.e. $S_d < 0$, the soil moisture reserve is replenished, hence increasing the amount of soil moisture available for use in the next 10-day interval. This is carried out as follows: If $S_d < 0$ then, $S_t = S_{t-1} - S_d \Delta t$. When available soil moisture (S) surpasses field capacity (S_{max}), deep percolation

(Q_{dp}) occurs: If $S_t > S_{max}$ then, $S_t = S_{max}$ and $Q_{dp}\Delta t = max(0, - S_d\Delta t - S_{max})$. However, the occurrence of deep percolation will depend on supplemental irrigation water application technology. In our case, deep percolation will be negligible since low-head drip irrigation water application technology was adopted. Otherwise deep percolation would only occur due to excess rainfall, and can be estimated on a daily soil moisture analysis and accumulated over the 10-day time step. It is expected that, except for initial 10-day intervals (i.e. at shallow rooting depths), all excess water is drained at each time step. Otherwise the soil moisture at the end of a cropping season should be equal to the moisture storage at the beginning of the next season (i.e. S_{t-1}).

Irrigation water management

The 10-day SIR was compared with water stored in the farm pond to ascertain whether the RHM system has adequate water to mitigate the dry spells. Seasonal farm pond water balance, dry spells and soil moisture analysis formed the basis of agro-hydrological evaluation of the on-farm RHM system. Water application methods, especially for irrigating kitchen gardens, were assessed through field survey. Due to the small quantities of water harvested and stored in the farm ponds, it was necessary to evaluate one of the efficient water application technologies that would use water judiciously. The impact of low-head drip irrigation technology was studied on a few farms with emphasis on crop performance and water utilization.

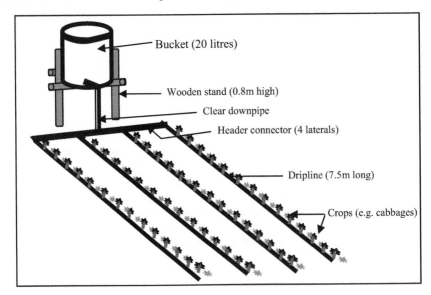

Fig. 4.5. Typical 20-litre low-head drip irrigation system mainly used for kitchen garden

The amount of water supplied by drip irrigation, i.e. the recommended 2-3 irrigation applications amounting to 40-60 l day^{-1} for a 15 m^2 area planted with 100 plants (Ngigi *et al.*, 2000; Ngigi, 2002b), was compared to the actual daily crop water requirement. A 20-litre bucket kit low-head irrigation system, as shown in Fig. 4.5, was used. The bucket is filled 2-3 times a day, early in the morning and late afternoon to avoid higher direct evaporation losses. The crop water requirement (water depth) was converted to volume for the irrigated area and compared with the amount of water supplied by the drip irrigation system. The use of drip irrigation to supply supplemental irrigation water requirement during the dry spells was also assessed with respect to the farm pond storage. Drip technology was considered due to its high efficiency, simplicity and limited quantity of water in the farm ponds.

4.2.6 Benefit-Cost Analysis

The analysis considered all the cost aspects related to the rainwater storage system, water application system and related farm inputs. The prevailing market price of vegetables (cabbages) was used to estimate direct income. However, although one crop was used in the analysis, farmers usually plant a variety of low and high value vegetables on their kitchen gardens.

4.3 Results and Discussion

4.3.1 On-farm Storage RHM Systems

The field evaluation revealed that on-farm RHM systems are common in central parts of *Laikipia* district. Sizes of the farm ponds range from 30-100 m^3 and catchment areas vary from 0.3-2 ha. The farm ponds harvested runoff from either natural catchment located adjacent to the ponds or from road/natural water courses/footpaths/cattle-tracks. Runoff was directed into the ponds by excavated ditches. The parameters of the two farm ponds studied are shown in Table 4.2.

Most of the evaluated farm ponds were not performing as intended in terms of harvesting and storing adequate runoff to meet the water demands (Ngigi, 2003a). The poor performance could mainly be attributed to high water losses, especially through evaporation and seepage. Moreover, poor water management also led to water losses, which means that harvested water only lasted for short periods of time after the rains.

Table 4.2. Parameters of the evaluated farm ponds in *Matanya*, *Laikipia* district

Farm pond parameters	Farm pond 1	Farm pond 2
Total volume (m^3)	32.0	40.0
Catchment area (ha)	0.5	0.4
Catchment gradient (%)	0.5-1.0	0.5-1.0
Maximum water surface area (m^2)	60.0	75.0
Top radius, r_1 (m)	4.4	4.9
Bottom radius, r_o (m)	1.0	1.1
Side slope (%)	36	33
Maximum pond depth (m)	1.2	1.3

The water losses were found to account on average to 30-50% of the stored water. Worse cases were reported in farms with sandy soils, where most of the water was lost almost immediately after the rains. Therefore, stored water was not adequate to meet the crop water requirements either to mitigate intra-seasonal drought or off-season irrigation. Moreover, water shortage was aggravated by low storage capacity (i.e. 30-50 m^3) of most farm ponds. Such low storage capacities are not adequate for full irrigation application, but can be used for supplemental irrigation, either to mitigate intra- and/or off-season dry spells. The poor performances of farm ponds may have led to inadequate maintenance or abandonment, and generally low adoption by other farmers.

Rainfall-runoff relationship

Runoff generation from the catchments ranged from 0-30% of the total rainfall depending on antecedent soil moisture conditions. The rainfall-runoff relationships of the study area shown in Fig. 4.6 indicate a threshold rainfall of 10-15 mm per storm, below which no runoff is generated. The low threshold rainfall is attributed to the clay soils predominant in the catchment and also the small catchment areas. The scatter is caused by different moisture conditions preceding rainfall events. Nevertheless, the rainfall generated enough runoff that filled most of the farm ponds during each rainfall season.

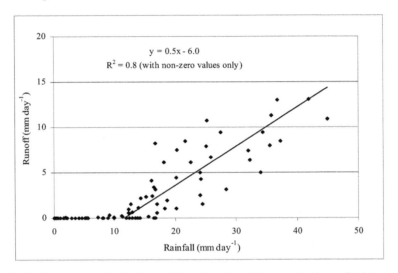

Fig. 4.6. Small catchments' rainfall-runoff relationship in *Matanya, Laikipia* district

4.3.2 Accumulated Runoff during the Rainy Season

The seasonal water balance analysis, presented in Table 4.3, was carried out for the seepage controlled farm pond 2 (lined ultra-violet resistant plastic material) during the short and long rains (2001-2003). The water balance was carried out with the intention of determining the amount of water stored at the end of the rainy season, and hence water withdrawal for irrigation was not included. This is in agreement with most farmers' experience where the stored water is used after the rainy season (off-season) and not during the rainy season (i.e. mitigation of dry spells is not yet being considered). Nevertheless, small farm ponds may fill up more than once within a rainy season, and hence their yield (flux, volume per time) can be larger than their storage capacity (i.e. stock, volume). This is an important consideration, especially where stored water is used to mitigate intra-seasonal dry spells.

Under extreme conditions, especially when intra-seasonal dry spells occur, supplemental irrigation may be necessary depending on crops growth stage and sensitivity to water stress. Table 4.3 shows that the seasonal rainfall can generate adequate runoff to fill most farm ponds in the study area whose storage capacity ranges from 30-50 m³. However, Table 4.3 shows that the available storage capacities are inadequate to store all the available runoff. Though the amount of water harvested seems directly proportional to seasonal rainfall totals, antecedent soil moisture conditions, rainfall distribution and intensity are also determinant factors. However, the rainfall occurrence in the study area follows a distinct pattern with seasonal variations mainly in the amounts.

Table 4.3. Farm pond 2 water balance analysis for the short and long rains in 2001-2003

Season	Rainfall (mm season^{-1})	Inflow		Outflow	Farm pond storage (m^3)
		Direct rainfall (m^3 season^{-1})	Runoff (m^3 season^{-1})	Evaporation (m^3 season^{-1})	
Short rains 2001	357	16	39	13	42
Long rains 2002	395	20	48	16	52
Short rains 2002	370	18	37	11	44
Long rains 2003	440	22	64	10	76

Water losses

Seepage and evaporation losses were found to vary from one farm pond to another due to size and side slopes of the farm ponds and spatial variation in soil characteristics. Even farm ponds on the same farm showed different results. The relationship between seepage and water depth for the farm ponds fitted on a power function due to their truncated cone shapes. The relationships between water depth and water losses (seepage and evaporation) show that water losses increase with water depths and exposed surface area. The evaporation and seepage losses ranged between 0.1-0.3 m^3 day^{-1} and 0.03-0.4 m^3 day^{-1} respectively (Thome, 2005) and accounted for 30-50% of the total harvested runoff. Evaporation rates in the study area are high ranging from 5-8 mm day^{-1} and since the farm ponds are not covered, more water was lost due to evaporation than seepage for farm ponds on clay soils. However, for farm ponds on sandy soils, where seepage rates is over 2 m^{3}day^{-1}, seepage losses accounted for more than 80% of total water losses.

Therefore, water losses can be reduced by lining the farm ponds (for example, with ultra-violet resistant plastic lining) and/or covering the ponds either by roofing with locally available materials or planting non-fruiting passion variety. The seepage rate also depends on textural composition of the underlying soils, method of construction and compaction, life of the farm pond and maintenance. It was reported in a few cases that seepage losses reduced over time as a result of siltation, which seemed to seal the surface of the farm ponds depending on the clay and silt contents of the sediments.

The hydrological evaluation of the farm ponds revealed that one of the challenges was how to reduce the seepage and evaporation water losses. High water losses have certainly contributed to the low technology adoption rate, and hence low agricultural production. Some farmers had abandoned their farm ponds claiming they were not useful as they only stored water during the wet season— when the crops do not require additional water. A few innovative farmers had tried several techniques to reduce the seepage without much success. However, ultra-violet resistant plastic lining is one of the promising cost-effective seepage control option.

The effect of ultra-violet resistant plastic lining was evaluated and the results were encouraging, in terms of reducing seepage water losses. The two lined farm ponds attest to this, as they were able to store water for longer duration. The results suggest that there are benefits associated with controlling water losses and improving irrigation water management. Reducing water losses would provide more water for the crops especially to meet water demands during the dry seasons, which sometimes coincide with critical growth stages. The benefits of reducing water losses were greater in areas with sandy soils. Therefore, lining the farm ponds would improve their performance and enhance their adoption. The overall results would be improved agricultural production, food security and rural livelihoods.

4.3.3 Occurrence of Dry Spells

Frequency analysis of occurrence of long term seasonal rainfall revealed the onset and cessation dates, length of the rainy season, occurrence and length of intra- and off-season dry spells. The onset windows for the long and short rains are 5-18 March and 7-18 October, respectively. The rainfall cessation windows are 5-28 May and 8-25 December for the long and short rains, respectively. The lengths of the long and short rainy seasons are 55-80 days and 62-73 days, respectively, which mean that the lengths of the rainy seasons are shorter than the growing periods for most crops grown in the area. This coupled with occurrence of intra-seasonal dry spells affect agricultural production. The results of analysis of occurrence of 10 and 15 days dry spells, with respect to different planting dates within the onset window (i.e. early planting (EP), optimal planting (OP) and late planting (LP)) are presented in Table 4.4. The analysis showed the probabilities of a dry spell exceeding 10 & 15 days are on average 77% & 41% and 57% & 31% for the long and short rains respectively during the critical crop growth stage. This justifies the use of a 10-day time interval in the soil moisture analysis.

Table 4.4. Occurrence (%) of a 10- and 15-days dry spells at different crop growth stages and planting dates

Crop growth stages	Long rains						Short rains					
	>10days			>15days			>10days			>15days		
	EP	OP	LP	EP	OP	LP	EP	OP	LP	EP	OP	LP
Establishment	11	17	17	0	6	0	12	12	12	0	0	0
Vegetative	11	11	11	0	0	0	12	6	6	0	0	0
Yield formation	67	78	85	28	45	50	47	59	65	18	35	41
Maturity	44	44	44	0	0	0	35	35	35	0	0	0

Projecting the crop growing period, which commence at the onset, over the rainy season showed the occurrence of periods of water stress, i.e. the longest dry spell during the growing period and short intra-seasonal dry spells (ISDS), as shown in Fig. 4.7. The critical dry spells seem to occur after rainfall cessations, i.e. off-season dry spell (ODS), which shows that optimal crop yields are not feasible without supplemental irrigation. The results in Table 4.4 show high occurrence of dry spells during the yield formation stage, which is sensitive for most crops. Early planting may reduce the severity of dry spells during the critical growth stage by 15-20%. Early planting would also reduce the SIR. However, there is no significant variation on the occurrence of dry spells due to early planting during crop establishment, vegetative and maturity growth stages.

4.3.4 Mitigation of Dry Spells

On-farm water ponds harvest runoff from small catchments for supplemental irrigation either during intra-seasonal dry spells or off-season dry spells which occur after rainfall cessation before crop maturity. The soil moisture balance analysis was carried out to determine the SIR of the crop—cabbage with 90 days growth period. The seasonal soil moisture balance and SIR for a crop planted at the optimal rainfall onset for short rains (2001 and 2002) and long rains (2002 and 2003) defined in Eq. (4.9) are presented in Table 4.5. The crop water requirement (E_c) value includes crop transpiration (T_c) and soil evaporation (E_s) i.e. $E_c = T_c + E_s$. The soil on the farms where detailed soil analysis was carried out was mainly dark grey clay with field capacity of 40-60% and available soil moisture of 100-200 mm m^{-1}. See Annex 1 for detailed computation of supplemental irrigation requirement for 90-day growing period cabbage in *Matanya* for the short rains 2002.

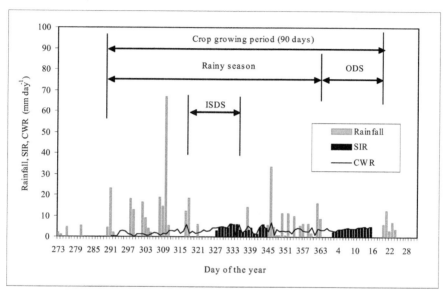

Note: *ISDS = intra-seasonal dry spell; ODS = off-seasonal dry spell; CWR = crop water requirement*

Fig. 4.7. Onset, cessation, dry spells, crop growing period and SIR for short rains 2002

Fig. 4.7 shows that the timing of supplemental irrigation does not coincide with the beginning of the dry spells as the crop first utilize soil moisture stored during the rainy days preceding the dry spells. From Table 4.5, the seasonal SIR values for the short and long rains are on average 103 mm and ranged between 85-125 mm. However, the SIR for the short rains (2002) is lower than that of the long rains (2002) despite the higher seasonal rainfall presented in Table 4.3. This is attributed to variations in rainfall distribution within the seasons and residual soil moisture from continental rains. Therefore, lower seasonal rainfall does not necessarily mean severe water stress.

Table 4.5. Seasonal soil moisture balance analysis and supplemental irrigation requirement

Season	E_c (mm season^{-1})	P_e (mm season^{-1})	S (mm season^{-1})	Q_{dp} (mm season^{-1})	Q_i (mm season^{-1})
Short rains 2001	332	163	79	8	98
Long rains 2002	334	174	47	12	125
Short rains 2002	328	170	72	19	105
Long rains 2003	336	208	54	11	85

Note: *E_c is crop water requirement; P_e is effective rainfall; S is available soil moisture storage; Q_{dp} is deep percolation; Q_i is SIR*

From the water balance analysis, moisture deficit increases towards the end of the growing period. The cabbage crop is harvested fresh and the crop requires more water towards maturity. The maturity period also coincides with the time when the rainfall amounts are low hence the increase in soil moisture deficit. The deficit represents the amount of irrigation water needed to meet the crop water requirements. The normal practice is to supply water up to field capacity, which could lead to deep percolation depending on water application methods. However, due to the limited amount of water harvested in farm ponds, the soil is wetted only at the root zone using low-head drip irrigation technology. This also ensures that water losses due to direct soil evaporation (E_s) is minimized to negligible values especially when the crop is well established. Although deep percolation is taken care by the effective rainfall, Table 4.5 indicates that deep percolation occurred when extremely heavy rainfall events were recorded.

Moreover, with limited water supply, considerations on crop selection and acreage are based on anticipated crop yields as affected by the extent to which crop water requirements are met by the available water supply. Where the amount of water stored is enough to meet the SIR, the soil is refilled with an amount equal to the moisture deficit in the given time interval. If the amount of water stored cannot meet the irrigation requirement, deficit irrigation is recommended. This can be accomplished by allowing the crop to undergo water stress during specific stages of the growing period when the crop is least affected by water deficit (Doorenbos and Kassam, 1979). It ensures enough water is available during the critical growth period, when water stress can affect crop yields.

The seasonal SIR analysis shows that the irrigation requirement increases as the planting dates move further away from the onset of rains. Within the onset window, early planting resulted to less severe dry spells occurrence, especially during the critical crop growth stages. During the rainy season, farmers rarely use supplemental irrigation as the rainfall is adequate, except when intra-seasonal dry spells occur. The seasonal amount of runoff harvested and stored in the farm ponds was found to be adequate to meet the SIR. Lowest irrigation requirement occurs when planting dates coincide with the onset of the long and short rains respectively. However, the irrigation requirement in the short rains is more than the long rains due to variations in seasonal rainfall. SIR is higher when crop growing period coincide with the driest periods of the year. By that time, the farm ponds are generally empty and hence crop production may not be feasible.

Irrigation water management

Farmers in the study area have been using hand watering application methods to water their kitchen gardens. However, due to the limited storage capacities and excessive water losses, this application method led to poor water management. The harvested water barely met the water requirements, which affected crop yields. The poor performance of the RHM systems affected the adoption of the technology. Most farmers were disappointed and some abandoned the technology all together. The agro-hydrological evaluation, therefore, focused on improving rainwater management by reducing water losses by introducing efficient water application system—drip irrigation technology.

The SIR in Table 4.4 shows that 50 m^3 farm ponds can adequately irrigate an average area of 480 m^2 ranging from 400-600 m^2. Considering irrigation water application efficiency and farmers' experience, this translates into at least an area of 300 m^2 that can effectively be irrigated. The evaluation of application rate of the low-head drip irrigation system revealed that the recommended 2-3 buckets (i.e. 40-60 l day^{-1}) for an area of 15 m^2 (Ngigi *et al.*, 2000) was adequate to supply the required amount of water. From a rough estimate of daily crop water requirements using the average evaporation rate of 6.5 mm day^{-1}, an irrigation unit of 15 m^2 (one bucket kit) will require 0.054 m^3 day^{-1} (i.e. 6.5 mm day^{-1} x 15 m^2 x 0.55). The factor 0.55 indicates the percentage of the soil profile of the cropped area wetted by drip irrigation, which ranges between 0.5-0.6 (Gathuma, 2000). Therefore, low-head drip irrigation system can adequately meet the daily crop water requirement. Lifting water using a hand pump complements water management, as it reduces water losses and risks of falling into the farm ponds.

4.3.5 Cost - Benefit Analysis of RHM System

The main method of water application used by farmers was hand-watering with buckets. The method usually leads to either under- or over-irrigation, non-uniform water application among plants and is prone to water wastage. High water use efficiency is a prerequisite where water supply is limited. This can be partially achieved through the use of highly efficient water application methods such as the low cost low-pressure drip irrigation systems which save on water and labour. A combination of RHM using low-head drip irrigation technology can improve water productivity.

Various sizes of low-head drip irrigation kits are available in the Kenyan market at prices ranging from US$ 15 for a 20-litre bucket kit to US$ 125 for the 200-litre mini-tank/drum kit. Low-head drip irrigation systems of intermediary sizes are also available to suit different farmer's preferences (Ngigi, 2002b). A farm pond with a 50 m^3 storage capacity can be used to supply supplemental irrigation water to a 300 m^2 (20m x 15m) kitchen garden, which can be covered by ten 40-litre bucket kits or two 200-litre mini-tank systems.

If irrigation is considered outside the rainy season, which may be advisable due to high demand of vegetables, a farmer can irrigate 150 m^2 with drip irrigation for 90 days. Under normal climatic conditions, the kitchen garden will require about 0.54 m^3 day^{-1} assuming no water losses, a 50 m^3 storage capacity farm pond would be adequate to supply irrigation water for 90 days after cessation of the rains, which would cover the growing season of most vegetables. Small amounts of water, and hence more irrigated areas, would be achieved if part of the growing season is within the rainy season. This implies more productivity with the same amount of irrigation water. The estimated investment cost for a farm pond RHM system for supplemental irrigation is outlined in Table 4.6.

With the existing agronomic practices, a 300 m^2 plot can yield 2,000 cabbages each weighing on average 1.5 kg. Assuming 10% field losses, at a price of US$ 0.15 per cabbage the seasonal gross return from the farm would be US$ 405, and hence a net revenue of US$ 305 per season. Higher value vegetables such as tomatoes and snow peas would even result in more returns. Considering two seasons per year (i.e. long and short rains), a farmer would have a net annual revenue of US$ 610. This is the maximum amount available for repayment of the loan that would be needed to construct the farm pond and install low-head drip irrigation system.

Table 4.6. Benefit-cost analysis of farm pond water management using drip irrigation system

Budget item	Cost (US $)
Construction of farm pond (20 man-days @ US$ 1.5)	30
Seepage control plastic lining sheet[8] 100 m^2 @ US$ 2.7 per m^2	270
Low-head drip irrigation system (i.e. for two 200-litres kits) @ US$ 125	250
Fencing and roofing	100
Total investment cost	*650*
Recurrent cost (labour and farm inputs) per season	100
Expected seasonal returns @ US$ 0.15 per kg of cabbage	405
Net benefit on investment per season	*305*

If a farmer takes a loan from a cooperative society at an interest rate of 10% or 12% per annum, the loan can be repaid within 3 years depending on the amortized cost of the loan in Table 4.7. The amortized cost will range from US$ 171-261 per year if the interest rate is 10% or from US$ 180-271 if the interest rate is 12% (see Table 4.7). For example, at 12% interest rate, the annual payment of US$ 271 means that a farmer can repay the loan and have substantial net revenue of US$ 339 from cabbage production during the repayment period. Therefore, on-farm storage RHM system is a viable investment in semi-arid areas, which experiences persistent crop failures, food shortages and poverty. The prices of vegetable increase during the dry seasons, which means that with RHM, farmers can manipulate crop production to coincide with high market prices. Farmers should be encouraged to take credit from existing institutions and invest in improved farm ponds and drip irrigation systems for vegetables production for the readily available market. Such investments would improve agricultural production and livelihoods of the poor farmers living in marginal lands.

Table 4.7. Loan repayment options at different interest rates and repayment periods

Present value (US$)	650					
Interest rate (%)	12			10		
Repayment period (years)	3	4	5	3	4	5
Amortized value (annual payment) (US$)	271	214	180	261	205	171
Monthly equivalent (US$)	23	18	15	22	17	14

4.4 Conclusions

The agro-hydrological evaluation entailed field survey, farm pond water balance analysis, soil moisture analysis, rainfall and dry spells analysis, calculation of SIR and benefit-cost analysis of cabbage production using drip irrigation system. Field survey revealed that the storage capacities of existing farm ponds ranged between 30-100 m^3, though most of them were on the lower range (30-50 m^3). Adequacy of farm ponds, in terms of size of the catchments, storage capacity and meeting SIR was also evaluated. The results revealed that considering rainfall characteristics, the catchment sizes can generate adequate runoff to meet SIR if water losses, especially on sandy soils, were controlled. Water balance analysis showed that evaporation and seepage losses account for 30-50% of the total seasonal water storage. Evaporation losses ranged from 0.1-0.3 m^3 day^{-1} and seepage losses from 0.03-0.4

---

[8] The unit cost of plastic lining varies with thickness; US$ 2.7, 3.1 and 3.9 for 0.8, 1.0 and 1.2 mm respectively.

m^3 day^{-1} on clay soils and more than 2 m^3 day^{-1} on sandy soils. However, despite the water losses, a 50 m^3 farm pond was found adequate to irrigate a kitchen garden of 300-600 m^2. Due to the limited amount of runoff, water use efficient drip irrigation was recommended instead of the current wasteful hand-watering method.

Soil moisture analysis was carried out on a 10-day time interval to determine the soil moisture depletion rates and hence SIR to bridge the dry spells. The soil moisture analysis covered the entire crop growing season, from rainfall onset to crop maturity. Some of the factors that affecting crop production in semi-arid areas are inadequate rainfall, occurrence of dry spells, shorter rainy seasons and poor timing of farm operations. Rainfall and dry spell analysis was carried out to determine whether farm ponds can reduce soil moisture deficit during the crop growing period. The dry spells analysis revealed that off-season dry spells, which occur after rainfall cessation, were longer and more severe than intra-seasonal dry spells. There is 80% probability of occurrence of dry spells exceeding 10 and 12 days during the long rains and short rains respectively. Therefore, supplemental irrigation is needed to mitigate the impacts of moisture deficit, especially during critical crop growth stages. The SIR for cabbages ranged from 85-125 mm.

The benefit-cost analysis revealed that incorporating drip irrigation with on-farm storage RHM system is economically viable, and a farmer can recover full cost of investments within 3 years while making substantial net revenue during the repayment period. However, despite economic viability and potential for improving agricultural productivity and livelihoods in semi-arid areas, their adoption is constrained by poor performance and socio-economic status of the poor farmers. The study focused on how to improve the hydrological performance of the farm ponds and associated economic benefits. The performance of the farm ponds can be improved by reducing seepage and evaporation water losses and adopting water use efficient drip irrigation systems. The side slopes and dimensions of the farm ponds should also be adjusted from 1:3 to 1:1 to reduce the exposed surface area. Trapezoidal shape should be adopted instead of truncated cone shape to ease construction and roofing to reduce evaporation. The farm ponds should also be provided with silt traps and spillways, and fenced to keep children and livestock from drowning. Incorporating appropriate agronomic practices such as soil and crop management would complement supplemental irrigation, and further improve agricultural productivity. Therefore, improved farm ponds provide one of the feasible options of reducing the impacts of water deficit that affect agricultural productivity in semi-arid environments in SSA.

Chapter 5

5.0 Agro-hydrological Assessment of In-situ RHM Systems[9]

5.1 Overview

RHM systems are increasingly being recognized as one of the strategies of improving food production, especially by small-scale farmers in semi-arid environments. The need to increase crop yields has led to adoption and up-scaling RHM systems such as conservation tillage and smallholder farm ponds. The chapter focuses on conservation tillage, which refers to soil tillage systems that minimize soil manipulation, increase moisture storage and reduce soil and water loss. It aims at conserving rainfall in-situ—where it falls on the cropland or pasture. Increased soil moisture storage reduces runoff, which lead to reduction in catchment water yield, which may reduce river flows and affect downstream water users. Hence the need to estimate the amount of rainwater retained on croplands and related hydrological impacts. However, it is not easy to estimate the amount of runoff—additional soil moisture—retained due to in-situ RHM systems such as conservation tillage practices.

The chapter presents an empirical approach based on the analysis of crop yield response to water. The results were compared with data taken from maize fields under traditional and conservation tillage systems in *Laikipia* district, *Ewaso Ng'iro* river basin in Kenya. Additional soil moisture retained in the root-zone enhances crop water availability leading to increased crop yields. Other water balance components such as deep percolation and long term soil moisture storage are negligible, while overall reduction in soil evaporation over the crop growth period is insignificant. The empirical approach is simple compared to other methods based on costly and time consuming field measurements. Field measurements showed that conservation tillage could increase soil moisture storage by 18-60%. Empirical approach yielded values within the same range, i.e. 15-40%. Therefore, empirical approach can be used to estimate the amount of rainwater retained on croplands, and hence proportion of runoff reduced from agricultural catchments.

The anticipated hydrological impacts related to up-scaling of RHM systems cannot be overemphasized (Ngigi, 2003b). The survival of rainfed agriculture in semi-arid environments relies on adoption of RHM systems such as on-farm storage ponds (Ngigi *et al.*, 2005a and 2005b) and in-situ rainwater conservation systems such as conservation tillage. In-situ RHM technology is distinct in that it does not include an external runoff generation area, but instead aims at conserving rainfall in

---

[9] *Based on:* Ngigi, S.N., J. Rockström and H.H.G. Savenije. (*forthcoming*). Assessment of rainwater retention in croplands due to conservation tillage and hydrological impacts in *Ewaso Ng'iro* river basin, Kenya. The paper has been submitted to the *Physics and Chemistry of the Earth, Special Edition*

the root zone—soil moisture for plant use. This reduce the amount of runoff generated, control soil erosion thus reducing the negative side effects of excess runoff. Other agronomic practices under this category include mulching, ridging, micro-catchment systems such as tied ridges/bunds, contour furrows/bunds and bench terraces. Soil structure improvements such as addition of manure would enhance the prospects of better yields.

In-situ systems are the simplest and cheapest RHM approaches and can be practiced in many farming systems. In-situ RHM systems, which are based on indigenous and traditional farming systems are common (Rockström, 2000a; LEISA, 1998 and Reij *et al.*, 1996). In a semi-arid context, however, especially with coarse-textured soil with high hydraulic conductivity, in-situ conservation may offer little or no guarantee against the poor rainfall distribution. Thus the risk of crop failure is only slightly lower than that without any measures. In such cases, other RHM technologies should be considered and incorporated in farming systems. The benefits of soil moisture conservation are more visible where soil fertility improvement measures are considered and incorporated (Barron, 2004). Other complementary agronomic practices such as timeliness of farm operation and early land preparation have added advantages.

Our focus is conservation tillage, which has been defined as any tillage sequence having the objective to minimize the loss of soil and water, and having an operational threshold of leaving at least 30% mulch or crop residue cover on the surface throughout the year (Benites *et al.*, 1998). However, in Kenya, farmers rarely achieve 30% crop residue cover. Therefore, with respect to small-scale farmers in a semi-arid environment, conservation tillage is re-defined as any tillage system that conserves water and soil while saving labour and traction needs (Rockström, 2000a). Conservation tillage aims at reversing a persistent trend in conventional farming systems of reduced infiltration due to compaction, soil crusting, hard (plough) pan formation and reduced water holding capacity due to oxidation of organic materials—due to excessive turning of the soil. From this perspective, conservation tillage qualifies as a form of RHM, where runoff is impeded and soil water is stored in the plant root zone (Rockström *et al.*, 1999). Unlike the conventional tillage systems, based on soil inversion which impedes soil infiltration and root penetration, conservation tillage covers a spectrum of non-inversion practices from reduced but deep tillage which aim to maximize infiltration, moisture storage and productivity, by minimizing runoff and evaporation, while saving energy and labour resulting in the reduction of production costs (Kaumbutho and Ochieng, 2001). However, reduced tillage could encourage weeds, which if uncontrolled, compete for the soil moisture.

The potential of conservation tillage is encouraging, especially with smallholder farmers who are already using animal drawn implements for their tillage operations, which are compatible with the conservation tillage tools. The common conservation tillage tools, which can be attached to the conventional animal drawn plough (e.g. *Victory* plough toolbar), are sub-soiler, ridger, *Magoye* ripper and tie-ridger. The compatibility of conservation tillage implements make them cheaper, as the farmers only need to buy the additional tools, which cost about KSh. 1,500 ($\approx$ US$ 20), and are locally available (Ngigi, 2003a). The most commonly practiced conservation tillage technologies are sub-soiling, ripping and ridging, in which effective rainfall may be supplemented by runoff diversion from external catchments such as roads, footpaths and natural watercourses. Sub-soiling, which entail use of a deep penetrating chisel-like tool, seems to have greater impact as it breaks the hard/plough pan associated with continuous use of the conventional mouldboard plough. The impact of conservation tillage is associated with increased soil moisture storage available for crop use, which consequently is reflected in improved yields.

Soil degradation in most semi-arid areas—soil erosion, soil fertility depletion, and soil compaction—can be attributed to surface runoff. Soil compaction arises from surface sealing and crusting, soil hardening and poor tillage practices. Continuous use of traditional tools such as the mouldboard plough has been found to cause impervious sub-surface hard (plough) pans, which impedes root penetration, infiltration and soil moisture storage capacity. The mouldboard plough often penetrates 8-15cm soil depths, depending on the soil condition and characteristics. However, conservation tillage equipments are able to penetrate soil depths ranging from 20-30cm (Ngigi, 2003a), which break the hard pan, increase soil moisture storage and reduce surface runoff.

Conservation tillage has several positive effects on water productivity (Rockström *et al.*, 2001) compared to traditional soil and water conservation systems. Besides enhancing infiltration and soil moisture storage, it improves timing of tillage operations and spatial distribution of soil moisture at field scale, which is crucial in semi-arid rainfed agriculture. In Kenya, promotion of animal drawn conservation tillage tools such as rippers, ridgers and sub-soilers among smallholder farmers in semi-arid *Machakos* and *Laikipia* districts has resulted in improved water productivity and crop yields (Ngigi, 2003a; Kihara, 2002 and Muni, 2002). There are many documented examples of successful conservation tillage practices in eastern and southern Africa, where crop yields have been increased through conservation of soil, water and nutrients and/or reduction in draught power needs (Rockström *et al.*, 1999 and Ngigi, 2003a).

The impacts of conservation tillage are encouraging its adoption by farmers, especially in semi-arid environments. However, increased adoption is likely to reduce runoff generation from agricultural catchments, which may have negative hydrological impacts downstream. Thus the need to quantify the proportion of rainwater retained on cropland and related hydrological impacts. Conventionally, surface runoff, soil moisture and drainage measurements mainly at runoff plots scale are used to estimate the water balance and thus the amount of rainwater harvested and utilized by the plants. However, these methods are tedious and expensive, as they require much time and costly equipments and instrumentation. The chapter presents an empirical approach of estimating the amount of rainwater retained in cropland and runoff reduction at field scale due to adoption of conservation tillage systems. The empirical approach is based on the crop yield response to water relationship. The approach is based on the assumption that incremental grain yields are attributed to additional soil moisture retained at the root zone for plant use.

5.2 Methodology

5.2.1 Description of Study Area

Kalalu is in *Daiga* location, Central division of *Laikipia* district and lies above 1900 m (see Fig. 5.1). The agro-climatic characteristic is semi-arid to semi-humid. The rainfall ranges between 600-1200 mm yr^{-1} with an average of 1024 mm yr^{-1}. Fig. 5.2 shows the distribution of long term monthly rainfall and evaporation. The main soil type is deep red clay soils (*ferric Luvisol*) with high water storage capacity (Liniger, 1991). Table 5.1 presents the soil characteristics. Soils have high water storage capacities of up to 250 mm within the rooting zone, which means about 50% of the water needed to grow maize can be stored in the soil with suitable in-situ RHM system. However, traditional tillage system limits optimization of soil moisture due to low cultivation depth and formation of hard pans associated with the use of conventional mouldboard animal drawn ploughs.

Table 5.1. Soil characteristics in *Kalalu, Laikipia* district

Depth (cm)	Textural class (%)				AWC (vol. %) (i.e. mm/100mm)	AWC (mm)
	Clay	*Silt*	*Sand*	*Organic matter*		
0 - 20	55	22	21	2	23.0	46
20 - 45	68	14	18	0	18.4	46
45 - 75	72	13	15	0	16.0	58
75 - 100	70	13	17	0	18.0	45
100 - 130	69	14	17	0	13.0	39
130 - 160	55	23	22	0	13.0	39
160 - 180	55	23	22	0	13.0	26

Note: AWC = available water content

In the last 4 decades, land-use has been changing from large-scale livestock to small-scale subsistence agricultural production systems. Most of the small-scale farmers migrated from the adjacent highly populated districts due to pressure on land. In SASE, agricultural production is constrained by unreliable and poorly distributed rainfall, which is characterized by periodic intra- and off-seasonal dry spells. The observed long-term rainfall and evaporation characteristics indicate water deficit (see Fig. 5.2).

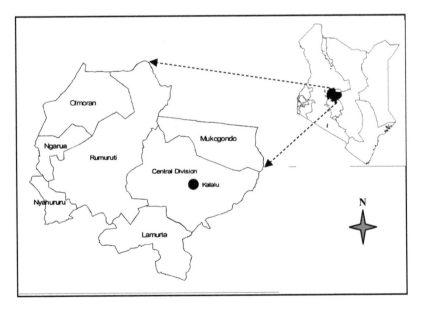

Fig. 5.1. Location of *Kalalu* in *Laikipia* district of Kenya

Inadequate soil nutrients also affect agricultural production. The solution lies on how to conserve and make optimal use of rainwater (Liniger, 1991) and maintenance of soil fertility. The farmers have taken up the challenge by adopting different RHM systems such as conservation tillage and farm ponds to address persistent water deficits and subsequent crop failures. Without adequate soil and water management, the agricultural potential cannot be realized. Ministry of Agriculture in collaboration with development partners is working with farmers to improve agricultural production through improved land and water management practices.

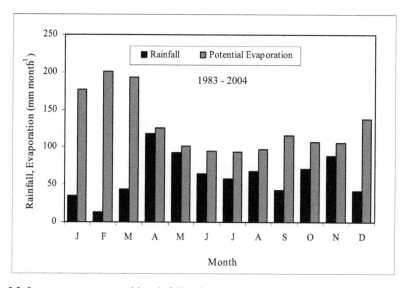

Fig. 5.2. Long term mean monthly rainfall and evaporation in *Kalalu, Laikipia* district

5.2.2 Crop Yield Response to Water

Evaluation of RHM systems in *Laikipia* district (Ngigi *et al.*, 2005b; Ngigi, 2003a and Kihara, 2002) revealed that conservation tillage is one of the promising soil moisture conservation practices used by farmers. Maize (*Zea mays*, L.) is one of the main crops grown by smallholder farmers in *Laikipia* district. The analysis was based on comparison of grain yields from farms under conservation tillage and traditional farming system. Traditional farming system here entails farm operations using conventional mouldboard plough and/or hand implements. Farms used in the analysis were assumed to have homogenous soil characteristics and similar agronomic practices such as land and crop management—soil fertility and pest control. Thus change in crop yields was due to additional soil moisture.

The amount of additional moisture attributable to conservation tillage can be estimated from crop yield response to water relationship adopted from Doorenbos and Kassam (1979) shown in Eq. (5.1). The relationship shows that crop yield decrease with increasing water stress. For many plants the actual crop transpiration T_a has a direct relationship with dry matter production or crop yield. When $T_a = T_m$, soil moisture is optimum, and hence crop produces optimum yields as no moisture stress exist. However, when $T_a < T_m$, the crop start to experience moisture deficit resulting to reduced crop yield i.e. less than Y_m. This forms the basis of crop yield prediction based on the level of soil moisture availability. Eq. (5.1) reflects the productive (transpiration) component, which directly influences crop yield.

$$\left(1 - \frac{Y_a}{Y_m}\right) = K_y\left(1 - \frac{T_a}{T_m}\right)$$

(5.1)

where; Y_a = actual grain yield (kg ha^{-1}), Y_m = maximum grain yield (kg ha^{-1}), T_a = actual crop transpiration (mm/season), T_m = maximum crop transpiration (mm/season), and K_y = yield response factor (i.e. effect of moisture stress on yield for the given crop).

Under rainfed agriculture, optimal crop growth conditions are rarely realized. This is due to occurrence of dry spells related to poor rainfall distribution over the growing period and poor rainfall partitioning (low infiltration and high surface runoff). Conservation tillage increases soil moisture storage, compared to traditional farming system, and hence improves crop yields. From the incremental grain yields, the additional amount of rainwater stored in the root zone and utilized by crops was determined. This provides an estimate of additional amount of rainwater retained in the root zone due to conservation tillage.

5.2.3 Empirical Estimation of Additional Soil Moisture

The additional soil moisture storage due to conservation tillage was estimated using maize yields from demonstration farms where field trials on conservation tillage have been carried out by the Kenya Conservation Tillage Initiative (KCTI) (Kihara, 2002; Kaumbutho and Mutua, 2002; Kaumbutho and Ochieng, 2001). Additional data was collected from some of the farmers who have continued with their own trials after the KCTI demonstration.

Eq. (5.2) was derived from Eq. (5.1) by taking T_m and Y_m as the crop transpiration and crop yield under conservation tillage (improved soil moisture condition), whilst T_a and Y_a as crop transpiration and crop yield under traditional tillage practices (low soil moisture condition). T_{TT} is the total seasonal crop transpiration under traditional tillage (i.e. $T_c = T_a$), which can either be estimated from soil water balance analysis or from Eq (5.1). The relationship between T_{CT} and T_{TT} is given in Eq. (5.3).

$$\left(1 - \frac{Y_{TT}}{Y_{CT}}\right) = K_y\left(1 - \frac{T_{TT}}{T_{CT}}\right)$$

(5.2)

$$T_{CT} = (1 + \theta_S)T_{TT}$$

(5.3)

where; Y_{TT} = grain yield under traditional tillage practices (kg ha^{-1}), Y_{CT} = grain yield under conservation tillage practices (kg ha^{-1}), T_{TT} = crop transpiration under traditional tillage (*TT*) practices (mm/season), T_{CT} = crop transpiration under conservation tillage (*CT*) practices (mm/season), and θ_S = additional soil moisture storage (%).

Eq. (5.4) was derived from Eqs. (5.1, 5.2 & 5.3) by making the following assumptions; (i) deep percolation is negligible as enhanced root development would capture soil moisture that may be lost below the root zone, (ii) change in soil moisture over the crop growing season is negligible, and (iii) the difference in soil evaporation under conservation and traditional tillage farms over the crop growing season is insignificant. However, conservation tillage may reduce soil evaporation by up to 30% at the early crop development stages due to crop residue left on the surface and reduced soil disturbances. Moreover, under excessive rainfall, deep percolation may occur under both tillage practices. Nevertheless, the assumptions would hold under normal circumstances and the additional soil moisture storage due to conservation tillage can be obtained from Eq. (5.4). The additional soil moisture storage is reflected as increased crop water use and hence the incremental yields.

$$\theta_s = \frac{\left(1 - \dfrac{Y_{TT}}{Y_{CT}}\right)}{K_y - \left(1 - \dfrac{Y_{TT}}{Y_{CT}}\right)} \tag{5.4}$$

Y_{TT} and Y_{CT} were obtained from actual crop yields from farms with traditional tillage and conservation tillage respectively in the given environment. K_y expresses the effect of soil moisture deficit on yield for a given crop and depends on the drought resistance of the species and its growth stage ($K_y < 1$, more resistant; $K_y > 1$, less resistant). K_y is based on crop growth stage, however, a weighted value was used for the growing season. The weighted yield response factor for maize, $K_y = 1.25$ (Doorenbos and Kassam, 1979) was adopted in the analysis. The additional soil moisture (i.e. $T_{CT} > T_{TT}$) would reduce the effect of water stress and hence improve yield (i.e. $Y_{CT} > Y_{TT}$). Thus an increase in T_c, results in yield increase, Y_{CT}, which can be attributed to additional soil moisture. The additional soil moisture storage indicates the extra amount of rainfall (i.e. increased infiltration and reduced surface runoff) that can be retained on croplands. Hence it estimates the percentage runoff reduction from agricultural catchments.

5.2.4 Surface Runoff Measurements

The impact of conservation tillage on surface runoff was assessed at runoff plot and field scale. At runoff plot scale, runoff was measured from plots with conservation and traditional tillage operations for two years (2002-2003), i.e. four cropping seasons. The size of the runoff plots[10] was 2m by 20m at a slope of 4%. The maize biomass and grain yields were also measured. Data from related past research (Liniger, 1991) supplemented the runoff plot measurements.

At farmers' field scale, surface runoff at the lower side of the farms was monitored for two years. The size of the runoff monitoring plots was 5m by 40m at slopes of 3-5%. Suitable farms, where monitoring was done easily at ponds fitted with a gauged staff were selected both conservation and traditional tillage treatments. Four farms for each treatment were monitored. Since the farmers' tillage operations and crop husbandry were not altered, farms with similar farming systems were considered. Land slopes and soil types were similar for the farms selected. Besides runoff, biomass and grain yields for maize were also recorded. Two years (2001-2002) data collected from previous experiments (Kihara, 2002 and Ngigi, 2003a) supplemented the results.

---

[10] The size of runoff plots ranges from 2-7m in width and 10-200m in length (Presbitero, 2003), while the standard USLE runoff plot is about 1.83m wide and 22.13m long, i.e. 40.5m².

Comparison between runoff data from plots and farmers' field was used to estimate runoff reduction that can be attributed to conservation tillage. Increase in biomass and grain yields was attributed to the additional soil moisture—rainwater retention on croplands. The percentage additional soil moisture obtained from empirical approach was compared with percentage runoff reduction from runoff plots and farmers' field measurements.

5.3 Results and Discussion

5.3.1 Impacts on Grain Yields

The results show that conservation tillage improves crop yields compared to traditional tillage under similar conditions. There was on average 30-150% increase in yields for beans, wheat and maize (Ngigi, 2003a and Kihara, 2002). Thus, by practicing conservation tillage, farmers are able to improve their crop yields, without necessarily adding any extra farm inputs. While beans in the area are normally affected by unpredictable intra-seasonal dry spells that occur during their growing stage, there was more moisture available in the fields under conservation tillage, which enhanced growth and yield improvement. Further yields increase may be achieved over the seasons due to improved soil structure.

Similarly, there was significant increase in biomass production (maize stovers and other crop residues, which are fed to livestock, grazed or left as mulch and green manure). Therefore, with conservation tillage, an ordinary farmer could attain or even exceed the optimal crop yields recommended by Jaetzold and Schmidt (1983). On average, farmers practicing conservation tillage were able to achieve at least 75% of the recommended optimal yields, especially for maize. Comparison between adjacent large-scale farms practicing mechanized conservation tillage, and small-scale farms using animal drawn conservation tillage techniques, revealed that a small-scale farmer can realize at least 60% of wheat yields attained by a large-scale farmer (Kihara, 2002). Conservation tillage also reduces the crop maturity period by 16% for maize and wheat and by 22% for potatoes (Kihara, 2002). According to Melesse *et al.* (2002), conservation tillage in Ethiopia was found to increase soil moisture and improve grain yield of maize by 22%. However, moisture conservation and yield improvement depends mainly on seasonal rainfall amount and distribution, soil characteristics and crop management. Summarized result on grain yields variations at farmers' field and runoff plots under conservation and traditional tillage practices are presented in Table 5.2 (see Annex 2 for details).

Table 5.2. Percentage increase in grain yields due to conservation tillage

Experimental site	Number of observations	Percent (%) increase in grain yield		
		Range	Mean	Std deviation
Farmers' field	16	20-50	28	10
Runoff plots	20	30-150	40	19

The effect of conservation tillage on grain yields is slightly higher at runoff plots than farmers' field scale. Early crop maturity reduces the risk of intra-seasonal dry spells which are common in *Laikipia* due to poor rainfall distribution, and helps remove crops from the fields early enough to pave way for land preparation for the next growing season. It also enables the farmers to use the animals when they are still healthy and strong. For most farmers they get an added advantage of being able to graze their livestock on the crop residues prior to cultivation for the next season. Early harvesting, also allows the farmer to sell their produce at optimal market prices and reduce post harvest losses.

Fig. 5.3 shows comparison between grain yields under conservation and traditional tillage practices. Some of the results indicated insignificant yield increase probably due to factors such as soil types, tillage equipments and agronomic practices (pest control, soil fertility, seasonal rainfall variations, etc.). Similarly, grain yield variations among different experimental farms could also be attributed to these factors. Soil fertility improvements, through application of fertilizers, enhanced realization of anticipated benefits—increased grain yields. However, most of these factors were controlled to eliminate their effects by carrying out the experiments under similar conditions except for different tillage practices.

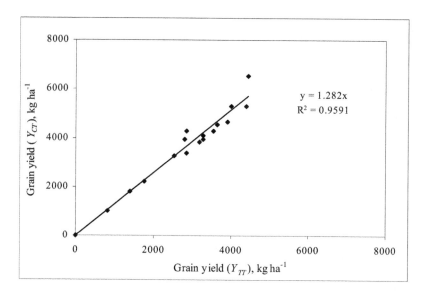

Fig. 5.3. Relationship between grain yields under conservation and traditional tillage farms

5.3.2 Additional Soil Moisture Storage

The additional soil moisture storage under conservation and traditional tillage practices at farmers' field and runoff plots are presented in Table 5.3. Field results revealed that conservation tillage leads on average to a 21% increase in soil moisture storage and plant water use (ranging from 15-40%) compared to traditional tillage practices. The additional soil moisture storage and hence increased crop water use is reflected in an average of 28% grain yields increase shown in Table 5.2, which range between 20-50%. The highest percentage additional soil moisture storage and use (i.e. 40%) was recorded on a farm where road runoff was diverted into the conservation tillage plot, thus supplementing the soil moisture and resulting in 50% grain yield increase. The additional soil moisture is attributed to increased infiltration and hence it can be used as an indication of percentage surface runoff reduction from agricultural catchments due to conservation tillage.

Table 5.3 shows that conservation tillage was found to improve soil moisture on average by 25-30% varying between 18-60%, with better results for higher rainfall and heavier storms (see sample data in Annex 3). Runoff plots showed a slightly higher percentage of grain yield and soil moisture increase compared to farmers' field. This could be attributed to spatial distribution of soil moisture, which is more significant at farmers' field compared to runoff plots. The results from runoff plots measurements showed that conservation tillage enhanced in-situ soil moisture conservation by trapping and storing rainfall as follows: 50-85%, 75-100% and 100% for heavy, medium and light storms respectively, compared to traditional tillage which captured and stored 25%, 50-60% and 75-100% respectively for similar storms.

Table 5.3. Additional soil moisture due to conservation tillage practices

Experimental site	Number of observations	Additional soil moisture (%)		
		Range	Mean	Std Deviation
Runoff plots	20	20-60	30	14
Farmers' field	16	18-50	25	16
Empirical approach	-	15-40	21	7

From the relationship in Fig. 5.3 (i.e. $Y_{CT} = 1.28Y_{TT}$), Eq. (5.4) estimates the additional soil moisture of 21%, which is due to conservation tillage. Therefore, 21% additional soil moisture leads to 28% increase in grain yield. The mean and the range of additional soil moisture computed from empirical approach is comparable to those obtained from field data as shown in Table 5.3 and Fig. 5.4. Thus, empirical approach can be used to estimate the percentage reduction of surface runoff from agricultural catchments due to conservation tillage practices.

Field data analysis and empirical approach shows that the relationship in Eq. (5.4) can hence be presented as shown in Eq. (5.5). Though, we had assumed that all the additional moisture will be used productively by the crops, some losses (evaporative and deep percolation) may occur. Field data indicate a negligible soil moisture loss of 2.6%—unproductive component of retained rainwater.

$$\theta_S = \frac{\Delta Y}{K_y} + \theta_L \tag{5.5}$$

where; ΔY = increase in grain yield (%), θ_s = additional soil moisture (%) and θ_L = non-productive soil moisture (%).

5.3.3 Assessment of Hydrological Impacts

The reduction in surface runoff is reflected in increased soil moisture storage and crop water use leading to improved grain yields. Quantification of reduced runoff from agricultural catchments due to adoption of conservation tillage, form the basis of assessing hydrological impacts downstream. The rainfall-runoff relationship from agricultural catchments under traditional and conservation tillage systems presented in Fig. 5.4, shows the reduction in surface runoff. Fig. 5.4 is based on farm/field scale runoff measurements similar to Fig. 3.2. The regression equations show the amount proportion of rainfall that ends up as runoff (i.e. runoff coefficient of 0.46 and 0.35 for traditional and conservation tillage respectively).

The difference indicates the amount of rainwater retained on agricultural catchments, which may subsequently lead to reduced river flows. Reduced river flows, though its occurrence during rainy seasons may not be significant, would affect downstream water users and impact negatively on the environment and natural ecosystems. The assessment of the hydrological impacts of up-scaling RHM systems in upper *Ewaso Ng'iro* river basin in Kenya (Ngigi, 2003b) cannot ignore the amount of rainwater that conservation tillage would retain in the farmers' fields. The additional soil moisture retained on the farms would reduce surface runoff from agricultural catchments, thus may to some extent, affect river flows and consequently downstream water availability. Hence, the results can be used to assess the hydrological impacts and determine the sustainable limit of up-scaling RHM in a river basin scale.

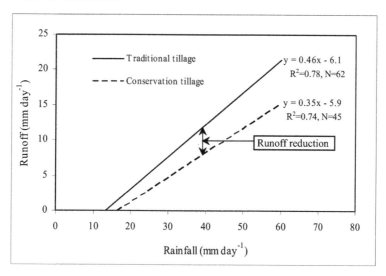

Fig 5.4. Rainfall-runoff relationship and runoff reduction due to conservation tillage

5.4 Conclusions

The analysis was based on grain yields for maize and runoff monitored from farmers' field and runoff plots under conservation and traditional tillage practices. The experimental fields had homogenous soil characteristics and similar agronomic practices such as soil and crop management. This was achieved by selecting adjacent farms and/or same farm practicing traditional and conservation tillage practices. The differences in grain yields were attributed to the additional soil moisture due to conservation tillage leading to increased crop water use.

Farmers' field evaluation revealed that adoption of conservation tillage could increase crop yields and reduce the crop growing period. Yield increase can be attributed to increased soil moisture and crop water use if plants through transpiration use the additional soil moisture in the root zone productively. However, other agronomic factors that affect crop growth may also contribute to yield differences. Some farms, under similar physical characteristics, recorded different grain yields, and hence human factors, especially farmer's experience and commitment to the research, also affected the results. This explains variations in grain yields on experimental farms.

Results from empirical approach compared well with field measurements. Therefore, empirical approach can be used to estimate the additional soil moisture retained on cropland due to conservation tillage. The empirical approach is simple and faster compared to field measurement methods, which are costly and time consuming. Additional soil moisture is translated into proportional surface runoff reduction from agricultural catchments. By multiplying the percentage additional soil moisture by rainfall, the amount of rainwater retained in agricultural catchments can be determined. Subsequently, reduction in surface runoff from croplands can be estimated.

The amount of rainwater—additional soil moisture storage—retained on croplands due to conservation tillage reduces surface runoff from agricultural catchment. This reduction in runoff may eventually lead to reduced river flows and hence negative hydrological and environmental impacts downstream, which would affect livelihoods and natural ecosystems. The knowledge of anticipated hydrological impacts forms the basis of quantifying the limits of up-scaling RHM technologies. The limit of up-scaling RHM is a prerequisite for formulating sustainable water resources management policies and strategies. Therefore, the information is important to policy makers, among other stakeholders, in their mission to improve agricultural production, food security and livelihoods in the vast and fragile semi-arid environments.

Chapter 6

6.0 Hydro-economic Analysis and Farmers' Investment Options [11]

6.1 Overview

Smallholder farmers in *Laikipia* district of Kenya, like their counterparts in water scarce semi-arid environments, are facing the challenge of improving agricultural productivity and livelihoods. A number of viable options are available, but high hydrological risks and low economic capability are discouraging the poor and risk-averse farmers. RHM is one of the promising options, whose impacts are unfortunately also affected by hydrological risks related to unreliable rainfall. The chapter presents a hydro-economic analysis of RHM systems with the aim of analyzing some of the factors that affect their adoption by smallholder farmers. Hydro-economic analysis included hydrological reliability of RHM systems, agro-hydrological risks and economic analysis. Agro-hydrological risk focused on dry spell and drought analysis, which affect soil moisture availability and hence crop production. Hydrological reliability assessed the ability of a RHM system to harvest and store adequate runoff to meet SIR to bridge dry spells and mitigates the impacts of persistent droughts. Economic analysis addressed benefit-cost analysis and profitability of RHM in terms of increasing crop production and stabilizing yields. The study was conducted in *Kalalu* and *Matanya*, which are in two different agro-climatic zones and represent land use changes in the recently settled areas of *Laikipia* district. The results provide a basis for farmers to make informed decisions on agricultural investments under hydrologic risks and uncertain production systems. RHM systems for supplemental irrigation were found to be an economically viable option for improving agricultural production and livelihoods of smallholder farmers in drought prone rural areas.

Agricultural investments, especially in the semi-arid environments are risky. Hence, most farmers are risk averse and reluctant to invest in new technologies that would improve agricultural production and livelihoods. The risk averseness can be attributed to poverty and bad experiences. Development agents tend to blame farmers of being slow and/or refusing to adopt new technologies without understanding the farmers' investment options and decision making process. Farmers, more than the so called development agents, are interested in improving their livelihoods, but are constrained by uncertainties and risks involved.

The challenge is: "how can farmers overcome the risks and uncertainties and invest in new technologies against the backdrop of their past experiences and

---

[11] *Based on:* Ngigi, S.N., H.H.G. Savenije, J. Rockstrom and C.K. Gachene. 2005. Hydro-economic evaluation of rainwater harvesting and management technologies: Farmers' investment options and risks in semi-arid *Laikipia* district of Kenya. *Physics and Chemistry of the Earth, 30:772-782*

limited options?" This leads to another pertinent question, "what needs to be done to cushion farmers against such risks to enhance adoption of promising technologies and improve agricultural productivity?" For instance, how do we convince farmers that RHM technologies are economically viable and worthy investing in? Researchers and development agents, more often confront farmers with only the positive aspect of a technology, and rarely highlight related risks and constraints. This has led to mixed fortunes, with some early adopting farmers, failing to realize the benefits thus discouraging others. Farmers are more concerned about, the failure of a technology and are more likely to remember failures rather than successes. The chapter presents opportunities and constraints related to adoption of RHM technologies by smallholder farmers in two climatically diverse zones of semi-arid *Laikipia* district.

While promoting RHM technologies as one of the feasible options for improving agricultural productivity in semi-arid environments, we need to understand the farmers' goals and decision making dilemma under risky and uncertain conditions presented in Fig. 6.1. These are some of the factors that influence farmers' decision making process. It is clear that RHM features predominantly as a viable option, but a farmer has other options too and the opportunity cost associated with the direction one may choose to follow to achieve the goals. Apparently, the farmers' goals and those of most development agents seem to converge despite the different perspectives on the best direction to take.

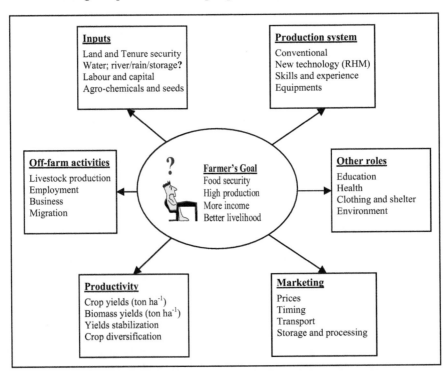

Fig. 6.1. Farmer's goal and decision-making dilemma under risks and uncertainties in SASE

The situation is complicated by the sectoral approach of development agents and government departments compared to the farmer's holistic approach. However, dialogue among stakeholders working towards improved livelihoods of the farming communities, could streamline the contributions and roles of different actors. Research should provide options for addressing constraints related to different options and assist farmers and others stakeholders to make informed decisions.

The chapter highlights the opportunities and constraints of adopting on-farm storage RHM systems for supplemental irrigation of maize and vegetables. Maize is the staple food in *Laikipia* and its production is periodically affected by inadequate and poorly distributed rainfall. Moreover, food security is associated with maize production. Although, irrigating maize may be uneconomical, Barron *et al.* (2003) showed that supplemental irrigation during flowering can substantially improve grain yields. Again, being the staple food, farmers would prefer to stabilize its production, even if only a small plot would be irrigated. Vegetable production is more economically viable due to higher returns and shorter growth period and high demand in semi-arid environments, which give it preference. The management of limited amounts of runoff harvested is improved by use of high efficient low-head drip irrigation systems.

The opportunities and constraints related to adoption of RHM systems are analyzed through hydrological risks assessment, agro-hydrological evaluation and economic analysis. The hydrological risks assessment focus on rainfall and drought frequency analysis, rainfall-runoff relationship and water balance analysis which determine the hydrological reliability of RHM systems in terms of storing adequate water. Agro-hydrological evaluation focuses on analysis of length and occurrence of dry spells, determination of SIR and adequacy and timeliness of RHM systems to meet SIR. The latter is more related to the risks of total crop failure or reduced yields due to occurrence of intra-seasonal dry spells and/or early rainfall cessation, i.e. rainfall season shorter than crop growing season.

The effect of rainfall onsets and variation in planting dates has also been assessed with respect to the magnitude and occurrence of dry spells. Economic analysis is broad, and in our case we have focused on benefit-cost analysis and improved livelihoods related to adoption of RHM technology. Therefore, our contributions in addressing the farmers' decision-making dilemma presented in Fig. 6.1 focus on inputs (making water available), production systems (adoption of RHM systems), productivity (improved crop and biomass yield, yield stabilization and crop diversity) and marketing (flexible timing of production and hence high prices and returns).

6.2 Methodology

6.2.1 Description of Study Area

The hydro-economic evaluation of RHM technologies on agricultural production was carried out in *Kalalu* and *Matanya* areas of *Laikipia* district (see Figs. 4.1 & 5.1). *Kalalu* and *Matanya* receive an average rainfall of 1024 mm yr^{-1} and 787 mm yr^{-1} respectively, whose distribution is shown in Fig. 6.2. They are classified as semi-humid and semi-arid zones with an elevation of 2020 and 1840m respectively. They are about 30km apart and represent two of the agro-climatic scenarios found in most medium potential areas of *Laikipia* district. Due to high altitude and relatively low rainfall, the soils have low mineral and organic matter content. However, the soils on wetter areas show a relatively high fertility but unfavourable soil water storage capacities (Desaules, 1986). *Kalalu* has mainly red clay soils (*ferric Luvisol*) whilst *Matanya* has dark clay soils with *vertic* properties (*verto-*

luvic Phaeozems) (Liniger, 1991). Soil characteristics are presented in Tables 4.1 and 5.1.

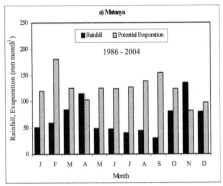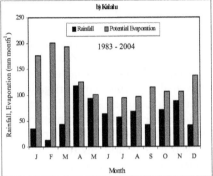

Fig. 6.2. Comparison of climatic characteristics of *Matanya* and *Kalalu* in *Laikipia* district

Recent land-use changes has resulted to land sub-division into uneconomically small plots (i.e. < 2ha). Agricultural production is affected by low rainfall and frequent crop failures. Thus drought coping strategies, both on- and off-farm, are important in order to maintain the land productivity and reduce the risk aversion of farmers. The marginal agricultural potential, small farm sizes, low capital availability for investments and increasing cost of living emphasize the importance of diversified production systems and alternative sources of income in order to ensure food security.

However, through external interventions and farmer's initiatives, a number of RHM systems have demonstrated their potential to address water scarcity and reduce periodic crop failures. Despite the significant impacts of RHM systems, one of the major challenges is how to convince the risk averse and economically constrained smallholder farmers to adopt and invest in RHM systems. The study aimed to achieve this by demonstrating how RHM systems can reduce hydrological risks and recover investment costs within a reasonable period. The effectiveness of RHM depends on a number of factors such as farmers' investments, rainfall reliability and distribution, water management, soil characteristics and types of crops. Farmers' investment options are constrained by hydrological risks, low land production and economic returns.

6.2.2 Hydrological Risks Assessment

Rainfall and drought analysis

Frequency analysis of 16 years observed rainfall records (1987-2002) for *Kalalu* and *Matanya* was carried out to determine the probability of occurrence and rainfall reliability in terms of meeting seasonal crop water demands. The rainfall analysis was also carried out for each year to determine the onset, cessation, rainfall duration and occurrence and magnitude of dry spells. Unlike in flood control, the design of RHM systems considers the probability ($P_{b(P)}$) of non-exceedence of a certain amount of rainfall, i.e. $P_{b(P)}(P < P_T)$ (Ngigi, 1996). Thus, system reliability (R_S), which indicates the expected performance of RHM systems in terms of runoff to be harvested, was determined from Eq. (6.1).

$$R_S(P \geq P_T) = 1 - P_{b(P)}(P < P_T)$$ (6.1)

where; P = rainfall amount (mm), and P_T = design rainfall over given return period, T (years).

The effect of staggering the planting dates within the rainfall onset window on occurrence of dry spells was also analyzed by considering early (start of onset window), optimal (actual onset) and late (end of onset window) planting. The onset of rainfall was determined by combining the FAO (1978) and Berger (1989) criteria. The FAO (1978) criteria define the rainfall onset as the time of the year when precipitation equals or exceeds $0.5E_o$, where E_o is potential evaporation. Berger (1989) qualifies this further by considering a 15 days period during which if the cumulative rainfall within 5 consecutive days is more than 20 mm, and the next 10 days also receive more than 20 mm, then the first day of the period marks the onset of the rainfall. Similarly, rainfall cessation is defined as the time when precipitation falls below $0.5E_o$, which may be extended for another 7 days to allow depletion of residual soil moisture. However, within the defined rainfall season, intra-seasonal dry spells occur, and if the crop growing period extends beyond the rainy season, off-season dry spells occur that may affect the crop at maturity stage. The dry spells affect crop production depending on their timing and magnitude with respect to crop growth stages and sensitivity to water stress.

Drought analysis was based on the observed long term seasonal and annual rainfall series. According to Ngigi (1996), the drought severity index (P_{di}) is determined from Eq. (6.2). The minimum water requirement for a crop under rainfed agriculture (i.e. 300 mm) has been used as the seasonal threshold value (P_D). From the drought severity index analysis, drought can be categorized as light, moderate or severe. Sharma (1994) used truncation levels similar to the drought index to categorize drought severity and suggested that RHM systems should be designed using parameters corresponding to moderate drought conditions since such designs would automatically take care of water deficit during severe droughts. One of the main design challenges of RHM systems is to increase their reliability—reduce chances of failure—without necessarily increasing the cost of the system. However, the probability of occurrence of drought, which is derived from the drought severity index, is one of the hydrological risks that affect adoption of RHM technologies in semi-arid environments.

$$P_{di} = \frac{P - P_D}{P_{sd}}$$ (6.2)

where; P = seasonal rainfall (mm), P_D = threshold rainfall (mm), and P_{sd} = standard deviation of seasonal rainfall (mm).

Dry spells analysis

The analysis of occurrence of rainfall seasons and dry spells (intra- and off-season) was carried out based on daily rainfall and evaporation data. The length of off-season dry spells was determined by superimposing actual crop water requirements with seasonal rainfall over the growth period commencing at the predetermined onset date. The probability of occurrence of a dry spell exceeding 10 and 15 days was determined for different planting dates. The adequacy of seasonal rainfall in meeting crop water requirement over the entire growth period was analyzed based on crop water requirements and soil moisture balance. When intra- and/or off-season dry spells occur, supplemental irrigation is applied to bridge the soil moisture deficit. In this case only adequate water for the most sensitive (critical) growing stage, in which water shortage would drastically affect yields, is applied—supplemental irrigation.

6.2.3 Agro-hydrological Evaluation

SIR for maize

RHM systems, in particular on-farm storage structures, are meant to reduce hydrological risks by bridging soil moisture deficit during intra- and/or off-season dry spells through supplemental irrigation. The SIR depends on the crop water requirement and soil moisture deficit. The timing and amount of SIR was from soil water balance analysis. The crop water requirement can be estimated from evaporation data. The potential evaporation (E_o) was estimated from *Class A Pan* evaporation data. A pan coefficient value of 0.8 was used to convert the pan evaporation (E_{pan}) to potential evaporation (E_o). The crop water requirement (E_c) for maize with 125 days growing period has been determined according to Doorenbos and Pruitt (1977) and Allen *et al.* (1998).

According to Ngigi *et al.* (2005a), the SIR can be computed by applying water balance approach to soil moisture balance analysis, which is based on the following assumptions:

- There is no groundwater contribution and deep percolation (Q_{dp}) below the root zone occurs when the soil moisture exceeds the soil field capacity.
- Planting is done at the onset of rainfall, and the initial soil moisture reserve on the planting day is taken to equal half the available soil moisture in the rooting depth.
- The rooting depth for maize is taken as 30, 60, 80 and 120 cm for initial (20 days), vegetative development (35 days), mid-season/yield formation (40 days) and late season/maturity (30 days) stages respectively.
- A crop coefficient (K_c) for maize of 0.40, 0.75, 1.15 and 0.60 for initial, vegetative development, mid-season and late season stages are used respectively.
- Drip irrigation with irrigation efficiency of 90% was adopted.

The soil moisture balance was carried out at a daily time step with the start of the first time step corresponding to the planting day (rainfall onset). For each time step, the soil moisture balance of the unsaturated zone was computed from Eq (6.3):

$$\frac{S_t - S_{t-1}}{\Delta t} = P_e - E_c \tag{6.3}$$

where; S_t = soil moisture storage per unit surface area at time, t (mm), S_{t-1} = soil moisture storage per unit surface area at time, t-1 (mm), Δt = time interval (day), P_e = effective rainfall (mm day^{-1}), and E_c = crop water requirement (mm day^{-1}).

If there is a soil moisture deficit (S_d) to meet the crop water requirement, i.e. S_d > 0, then soil moisture is withdrawn from the root zone storage accrued during the previous time step (S_{t-1}). If $S_d \leq S_{t-1}$ then there is enough soil moisture available to meet the deficit and hence $Q_i = 0$; else, moisture stored in soil cannot meet the crop water requirement and supplemental irrigation is required. SIR is calculated from Eq. (6.4):

$$Q_i = \max \left(S_d - \frac{S_t - S_{t-1}}{\Delta t}, 0 \right) \tag{6.4}$$

In case of surplus moisture $(S_d < 0)$, the soil moisture reserve is replenished, hence increasing the amount of soil moisture available for use in the next time step. When available soil moisture (S) surpasses field capacity (S_{max}), deep percolation (Q_{dp}) occurs: If $S_t > S_{max}$ then, $Q_{dp} = S_t - S_{max}$ and subsequently, $S_t = S_{max}$. However, the occurrence of deep percolation will depend on supplemental irrigation application technology. In our case, deep percolation will be negligible since low-head drip irrigation water application technology will be adopted. Otherwise deep percolation would only occur due to excess rainfall, and can be estimated on a daily soil moisture analysis. It is expected that, except for initial crop growth stages (i.e. at shallow rooting depths), any excess water is drained at each time step.

The soil moisture balance analysis in Eq. (6.3) depends on estimation of crop water requirement and effective rainfall. Effective rainfall (P_e) is the proportion of rainfall (P) that is either directly, as soil moisture stored in the root zone, or indirectly, as surface runoff that is stored in reservoirs and applied through irrigation, available for and used by the plants use (Ngigi *et al.*, 2005a). Therefore, P_e incorporates surface runoff, interception and soil moisture storage (infiltration) within the root zone. Part of deep percolation (Q_{dp}) that can be utilized by the plants through capillary rise is also included. However, under extreme high rainfall, not all deep percolation can be accounted for by effective rainfall. P_e is estimated for our conditions using empirical Eqs. (6.5 & 6.6) (Dastane, 1974; FAO and WMO, *Undated*). The following equations show that P_e is proportional to P which explains the higher coefficient for rainfall above 2.33 mm day^{-1}.

$P_e = 0.6P - 0.33$ for $P < 2.33$ mm day^{-1} (6.5)

$P_e = 0.8P - 0.80$ for $P > 2.33$ mm day^{-1} (6.6)

Reliability of RHM systems

The reliability of RHM systems is based on the runoff volume harvested, storage period and crop water requirement. Runoff generation from the catchments ranged from 0-30% of the total rainfall depending on antecedent soil moisture conditions (Ngigi *et al.*, 2005a; Thome, 2005). Fig. 4.3 shows the rainfall-runoff relationships of the study area, which indicate a threshold rainfall of 10-15 mm per storm, below which no runoff can be generated. The low threshold rainfall can be attributed to the clay soils predominant in the catchment and the small catchment areas. The scatter is caused by different soil moisture conditions preceding rainfall events. The seasonal rainfall generates enough runoff to fill most of the farm ponds, whose storage capacity lies between 30-100 m^3 (an average of 50 m^3) (Ngigi *et al.*, 2005a). However, the reliability of the storage systems is reduced by high seepage and evaporation losses, which on average accounts for 30-50% of the total water stored (Thome, 2005).

Simulation of seasonal runoff generated from the 16 years of rainfall records was carried out to determine the reliability of a RHM system with a catchment size of 0.5 ha and cropped area of 0.2 ha in meeting the SIR. This is achieved by comparing the amount of runoff generated and SIR over the crop growing season. The amount of runoff generated was estimated from the rainfall-runoff relationship shown in Fig. 6.3. The number of cropping seasons when the RHM system does not meet irrigation water demand and storage capacity to store the generated runoff indicates the probability of system failure. The number of cropping seasons (as percentage of total number of seasons) when the storage capacity and crop water demands are equaled or exceeded indicates system success, and hence its reliability. Thus the probability of the RHM system to meet the supplemental irrigation water requirements is taken as its reliability.

In an attempt to reduce the hydrologic risks, optimal RHM systems have been designed for the two study areas at 80% seasonal rainfall amount reliability level. This is achieved by fixing the cropped area at 0.2 ha and iteratively determining the optimal catchment size and storage capacity that meet the SIR. Comparison between the optimal RHM system and existing farmers' systems has been carried out to determine the contributions of non-optimal designs to hydrological risks.

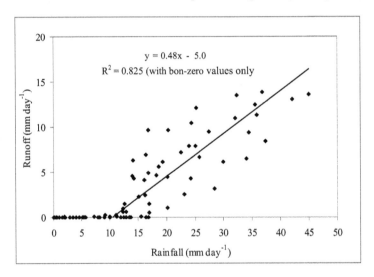

Fig. 6.3. Field scale rainfall-runoff relationship in *Kalalu, Laikipia* district, Kenya

6.2.4 Economic Analysis

Improvement of rainfed agriculture through adoption of RHM technologies is an investment where a farmer expends capital resources into a production system expecting to realize benefits over an extended period of time (Gittinger, 1972). To determine the economic productivity of adopting RHM, a simple benefit-cost analysis should suffice as one of the analytical methods of showing a farmer, in terms of social benefits, what he/she stand to gain or loose by deciding either to adopt or not to adopt the technology. A comparison between conventional rainfed farming system and RHM system was done to show the benefits associated with improved farming methods.

The analysis considered all the cost aspects related to the rainwater storage system, water application system and related farm inputs. The prevailing market price of dry maize grains was used in the benefit-cost analysis. However, although maize, which is the staple crop in the study area, was used for the analysis, farmers usually plant a variety of low and high value crops, hence incorporating crop diversity would even indicate higher returns. The water application system considered was a 200-litre low-head drip irrigation system which can irrigate up to 0.1 ha of maize. This means two drip irrigation kits for the 0.2 ha cropped area considered in the analysis. The acreage of cropped area under supplemental irrigation was based on the storage capacity of the existing RHM systems and cost of bigger systems which most poor farmers would not afford.

6.3 Results and Discussion

6.3.1 Rainfall and Drought Characteristics

Although *Kalalu* and *Matanya* have moderate annual rainfall totals, persistent agricultural droughts are a common feature as shown by the results of drought analysis presented in Table 6.1. Moreover, the probability of occurrence of seasonal rainfall lower than the long term average rainfall is 60% for both long and short rainfall seasons. This means that without RHM, there is only a 40% chance that farmers would produce some harvest. Thus adoption of RHM technologies would increase the chance of producing a good harvest. The results show that there is more than 50% probability of occurrence of moderate and severe droughts in semi-arid *Laikipia* district, which drastically reduces agricultural production and farmers' incomes. The probability of occurrence is even higher in drier areas such as *Matanya* with over 70% occurrence compared with semi-humid areas such as *Kalalu*. Flurry (1987) found rainfall reliability factors of 0.33 and 0.5 for *Matanya* and *Kalalu* respectively, which indicate the probability of receiving adequate rainfall.

Table 6.1. Probability (%) of drought occurrence in *Laikipia* district in the period 1987-2002

Drought severity	Long rains season		Short rains season	
	Kalalu	*Matanya*	*Kalalu*	*Matanya*
Light drought (-0.19 < P_{di} < -0.49)	18	33	8	30
Moderate drought (-0.50 < P_{di} < -0.79)	9	25	8	10
Severe drought ($P_{di} \geq$ -0.80)	42	83	50	73
Long term mean seasonal rainfall (mm)	282	209	201	295

In *Kalalu* the short rainfall seasons (October - December) have lower rainfall amounts than the long rainfall seasons (March - May), while in *Matanya* the short rainfall seasons have higher and more reliable than the long rainfall seasons. In the event of severe seasonal drought, crop failure or meager yields are expected. RHM systems solely rely on rainfall occurrence and system failure would be inevitable in the event of severe seasonal droughts. Therefore, RHM technologies may not entirely mitigate the impacts of persistent droughts but they reduce their effects by improving and/or stabilizing crop yields. The effectiveness of RHM systems is normally affected by the low storage capacities, which is constrained by high construction cost and farmers' unwillingness to take risk due to unreliability of the rainfall. Farmers are aware that higher investments in RHM would reduce the impacts of periodic droughts, but their low incomes and the related high risks hinder investments.

6.3.2 Rainfall Seasons and Dry Spells

Analysis of long term seasonal rainfall in *Matanya* indicated the onset and cessation dates, length of the rainy seasons and occurrence and length of intra- and off-season dry spells. The onset windows for the long and short rainfall seasons are 5-18 March and 7-18 October respectively. The rainfall cessation windows are 5-28 May and 8-25 December for the long and short rainfall seasons respectively. This shows high variations and unreliability of rainfall in terms of length and occurrence of rainfall seasons, which are defined by onset and cessation dates. The lengths of the long and short rainfall seasons are 55-90 days and 62-85 days respectively. This means that the lengths of the rainy seasons are shorter than growing periods for most crops grown in the study area. This coupled with occurrence of frequent intra-seasonal dry spells affect overall agricultural production. However, despite shorter rainfall seasons, crop survival in *Laikipia* district is enhanced by continental rains, which occur in between the two main rainfall seasons, i.e. June - September.

Projecting the crop water requirement (E_c) which commences at the rainfall onset and ends at the crop physiological maturity, over the rainfall season showed the occurrence of periods of water stress—intra-seasonal dry spells (ISDS) and off-season dry spells (ODS) as shown in Fig. 6.4. The critical dry spells occur during the second ISDS, which corresponds to yield formation stage and hence optimal crop yield is not feasible without supplemental irrigation. Thus the results show high occurrence of dry spells during the sensitive crop growth stage.

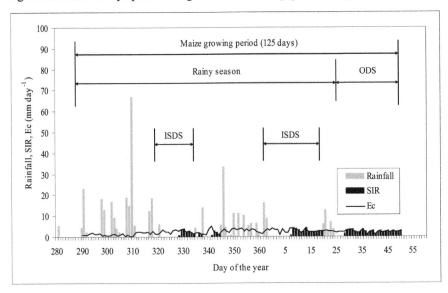

Fig. 6.4. Short rainfall season, crop growing period and dry spells for *Matanya* in 2002

Table 6.2 shows that early planting would minimize the severity of dry spells during the critical growth stage by 15-25% and also reduce the SIR. However, there is a 5-15% variation on the occurrence of dry spells due to early planting during crop establishment, vegetative and maturity growth stages.

Table 6.2. Probability (%) of dry spells at different crop growth stages and planting dates in *Matanya*

Crop growth stage	Long rains season						Short rains season					
	10 days dry spell			15 days dry spell			10 days dry spell			15 days dry spell		
	EP	OP	LP	EP	OP	LP	EP	OP	LP	EP	OP	LP
Germination	0	11	16	0	6	11	6	11	14	0	6	8
Vegetative	34	39	44	11	17	22	17	30	39	0	6	11
Yield formation	57	70	75	48	61	67	54	65	69	22	39	47
Maturity	46	57	61	26	34	40	28	36	42	20	32	35

Note: EP = Early Planting (at the lower quartile of the onset window); OP = Optimal Planting (at the median of the onset window); and LP = Late Planting (at the upper quartile of the onset window).

Fig. 6.5 shows that early planting has more impacts on reducing the severity of dry spells during the short rainfall seasons than the long rainfall seasons, due to better rainfall distribution. It is a common practice for farmers to plant when the rainfall begins regardless of whether it is within the onset window, which sometimes lead to seedlings wilting after germination, especially when false onsets occur. However, early planting within the onset window have better results. Late planting has no significant effect on the occurrence of intra-seasonal dry spells compared to optimal planting, though it lead to shorter rainfall season and reduce moisture availability and hence high SIR. Fig. 6.5 suggests that planting at the optimal onset in a particular season may not necessarily yield optimal results.

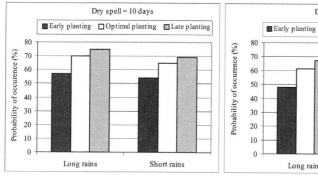
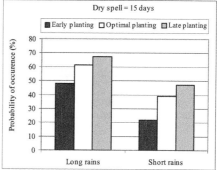

Fig. 6.5. Dry spells during crop yield formation stage for different planting dates for *Matanya* in 2002

The variation of rainfall onset window present another farmers' dilemma—when to plant to reduce the impacts of dry spells and ensure better crop yields. The uncertainty of rainfall onset adds to the number of hydrological risks encountered by the smallholder farmers in semi-arid environments. Therefore, there is need for adaptive and participatory action research to convince the farmers that RHM can reduce some of the risks and uncertainties involved in rainfed agriculture. The agro-hydrological analysis highlights a number of risks and decisions farmers have to make to improve agricultural productivity under uncertain situations related to unreliable weather conditions. Thus, one of our obligations and challenges as researchers is to develop decision support systems that will help the farmers reduce the risks and invest in promising RHM technologies to improve agricultural productivity and rural livelihoods in semi-arid environments. However, such decisions are often inhibited by inadequate resources and viable options to consider in decision-making process. One of the researchers' challenges is how to communicate such useful information to subsistence farmers. This challenge was addressed through participatory action research, where the study was carried out on farmers' field with full participation of the farmers. The farmers were encouraged to continue monitoring the performance of their RHM systems and to allow neighbouring farmers learn from the research sites.

6.3.3 Reliability of RHM Systems

The reliability of a RHM system shows its capacity to harvest adequate runoff to meet the SIR. Storage of more runoff means that higher RHM system reliability can be achieved, which may even lead to increased acreage under supplemental irrigation. The evaluation of reliability of existing RHM systems revealed that the storage capacity of most farm ponds was inadequate in storing the runoff generated by the catchment areas. The percentage of the storage capacity over the generated runoff volume gives an indication of the system capacity to harvesting and storing adequate volume, especially where runoff generating catchment is not limiting.

However, not all the generated runoff is utilized by the crops, and hence the reliability of RHM systems is expressed as the percentage of time the SIR is met by the stored water. Table 6.3 shows the percentage of the generated runoff that is harvested by farm ponds and also their reliability in terms of meeting the SIR. RHM systems in the semi-arid *Matanya* harvest a higher percentage of generated runoff but they have low reliability compared to the semi-humid *Kalalu* due to low seasonal rainfall, i.e. less runoff is generated for the same storage capacity. The low system reliability in *Matanya* is due to higher crop water requirement and lower amount of runoff stored—higher evaporation and lower rainfall. The higher runoff generated in *Kalalu* also requires bigger storage capacity and the probability of overtopping the 50 m^3 storage capacity is higher, which accounts for the lower percentage of generated runoff that can be harvested.

Table 6.3. Amount of generated runoff stored and reliability of RHM systems to meet SIR

Location	Long rains season		Short rains season	
	Runoff stored (%)	System reliability (%)	Runoff stored (%)	System reliability (%)
Kalalu	27	80	40	73
Matanya	53	67	47	69

In general, though reliability of RHM systems to meet SIR is greater than 60% in both areas, the 50 m^3 storage capacity used in the evaluation is not sufficient to store the runoff generated from the available catchments as indicated by low storage capacity reliabilities (30 - 50%). Rainfall in the areas occurs in large storms and consequently there are occasions when runoff generated from a single storm is enough to fill and even overtop the farm ponds. In spite of the cost implications, higher storage capacities would increase runoff storage, reduce runoff losses and hence increase the overall reliability of RHM systems. This would also increase acreage of crop under supplemental irrigation, crop yields and agricultural productivity in semi-arid environments.

To improve the overall reliability, optimal RHM systems should be designed and developed. In this regard, optimal design parameters of RHM systems in the two areas at 80% reliability of seasonal rainfall amount were determined as shown in Table 6.4. The system reliability depends on the rainfall amount and distribution, availability of catchment area to generate the required amount of runoff, storage capacity of farm ponds and the SIR. The higher the seasonal rainfall, the lower the designed catchment area and storage capacity. The highest seasonal rainfall amounts are experienced during the long rainfall seasons in *Kalalu* and the short rainfall seasons in *Matanya*. From the optimal design parameters presented in Table 6.4, it can be deduced that most of the existing RHM systems are not adequately designed to store adequate runoff to meet SIR.

Table 6.4. Optimal design parameters for RHM systems at 80% reliability of rainfall amount

Location	Long rains season		Short rains season	
	Storage capacity (m^3)	Catchment area (ha)	Storage capacity (m^3)	Catchment area (ha)
Kalalu	75	0.28	85	0.30
Matanya	130	0.50	95	0.35

Despite the small land holdings, availability of catchments to generate adequate runoff is not a limiting factor. There is vast area that has not been settled, and is used for grazing, which generates adequate runoff that can be diverted and stored in farm ponds by farmers in adjacent lands. There is also a high potential for diverting runoff concentrated by road/footpath/cattle track drainage systems that would complement natural catchments. Moreover, prevalent of high intensity rainfall in semi-arid areas leads to generation of high runoff even from smaller catchment areas. Optimal designs of RHM systems would improve overall system reliability and hence reduce the hydrological risks and crop failures. Nevertheless, the challenge is to convince the poor and risk-averse smallholder farmers that they can get more benefits by improving their RHM systems, which would translate to higher investments. Unfortunately, failure of poorly designed RHM systems has been found to discourage the would-be adopters, and hence the need to promote optimal systems that would be cost effective and realize higher economic returns.

6.3.4 Benefit-Cost Analysis

The driving force for the adoption and investments in RHM technologies is improved agricultural production, which leads to food security, income generation, poverty reduction and better livelihoods. Food self-sufficiency is the primary goal for most subsistence farmers. Maize, which is the staple food of smallholder farmers in *Laikipia* district, was used in the benefit-cost analysis. However, high value horticultural crops such vegetables would give higher returns and potential of investing in RHM technologies. The benefit-cost analysis was based on actual grain yields recorded by the Ministry of Agriculture (Kihara and Ng'ethe, 1999) from conventional and RHM systems as shown in Table 6.5.

Table 6.5. Comparison of seasonal maize yields in *Kalalu* and *Matanya, Laikipia* district

Grain yields (kg ha^{-1})	Long rains season		Short rains season	
	Kalalu	*Matanya*	*Kalalu*	*Matanya*
Conventional system	1,620	450	900	1,080
RHM system	2,250	630	1,260	1,530

Table 6.5 shows that RHM systems can improve grain yields—by increasing soil moisture availability over the crop growing season. On average the seasonal grain yield under conventional system in parts of *Laikipia* district is 1,000 kg ha^{-1}. This can be improved by integrating rainfed and irrigated agriculture through supplemental irrigation and improved agronomic practices.

Studies in *Machakos* district has shown an average yield increase of 38% can be achieved through supplemental irrigation and proper agronomic practices like fertilizer use (Rockström *et al.*, 2001). Soil moisture conservation technologies such as mulching have been found to increase grain yields in the study areas by 3-4 times (Liniger, 1991 and Liniger *et al.*, 1998). Therefore, the grain yields given in Table 6.5 can be further increased to 4,000-5,000 kg ha^{-1}, with improved supplemental irrigation water management (i.e. use of low cost drip irrigation system) and proper agronomic practices. Table 6.6 presents the results of benefit-cost analysis for a farmer who has adopted an improved farming system—RHM system comprising of a 50 m^3 farm pond and a low-head drip irrigation system for a 0.2 ha plot of maize compared with a conventional system without RHM system.

Table 6.6. Benefit-cost analysis of improved farming (RHM) and conventional systems

Description of cost item	Conventional system	RHM system
Construction work (20 man-days @ US$ 1.5)	-	30
Seepage control plastic 100 m^2 @ US$ 2.7 per m^2	-	270
Farm pond accessories (fencing and roofing)	-	100
Low-head drip irrigation system @ US$ 125	-	250
Total investment cost	-	*650*
Recurrent cost (labour and inputs) per season	150	200
Seasonal returns per ha @ US$ 0.25 per kg	250	450
Net benefit on investment per season	*100*	*250*

Table 6.6 shows that the expected seasonal net benefits are US$ 100 and US$ 250 for conventional system and RHM system respectively, which gives a net benefit of US$ 150 and a pay back period of about 4 (650/150) seasons. However, since a larger farm pond is more economical, the investment cost can be recovered in one year with a 100 m^3 farm pond. Thus, although the initial investment cost is high, which is one of the factors that discourage farmers' investments, the technology is economically viable. Moreover, investments in RHM system can double the farmer's agricultural income, and hence improve their livelihoods. Improved water management, soil fertility, crop husbandry and high value crops (as shown in section 4.2.5) would also lead to higher crop yields and incomes. Therefore, despite the relatively high investment cost and hydrological risks that inhibit many poor farmers, a well designed and managed RHM system can get smallholder farmers in semi-arid environments out of the vicious poverty cycle.

RHM has been found to reduce poverty in *Sipili* area of *Laikipia* district (Mbugua, 1999 and Kiggundu, 1999). The risk averseness and inadequate resources are some of the main factors inhibiting investments towards improving agricultural production and rural livelihoods. Therefore, improving agricultural production depends on reducing agro-hydrological and improving economic status of the farmers. Technically, hydrological risks can be reduced through optimal designs of RHM systems; whilst economic risks can be addressed by provision of credit facilities to improve investments in agricultural production. The loan repayment options, interest rates and repayment periods are presented in Table 4.7. However, the challenge still remains how to convince the risk-averse farmers to invest in improved agricultural systems.

6.4 Conclusions

Hydro-economic evaluation was based on the analysis of agro-hydrological risks and economic factors that affect investments in agricultural production systems that would enhance food security and improve rural livelihoods in semi-arid environments. Occurrence of persistent agricultural droughts is one of the limiting factors inhibiting agricultural development. The 60% probability of occurrence of below average rainfall and 50% probability of occurrence of droughts that affect crop yields depict a desperate situation. However, the drought severity is more pronounced in the more arid environment. Persistent drought aggravated by occurrence of intra- and off-seasonal dry spells (crop growth periods shorter than rainfall seasons) are the main hydrological risks that affect crop production and hence investments.

RHM systems can reduce the impacts of droughts, but their adoption is affected by high investment costs, economic status of the farmers, farmers' risk-averseness, inadequate design and poor water management. The reliability of RHM systems can be improved by increasing catchment area and storage capacity. The existing RHM systems have higher reliability in terms of meeting SIR and lower reliability in terms of their capacity to store generated runoff. The 30-50 m^3 storage capacity of many farm ponds was found to be inadequate at 80% rainfall reliability. However, despite the slightly higher cost associated with bigger storage capacities, the expected benefits—increased crop production and incomes—are encouraging. Improved water management through incorporation of low-head drip irrigation technology can improve the reliability of RHM systems and encourage farmers to increase their acreage under supplemental irrigation.

Benefit-cost analysis revealed that a farmer can recover his/her investment costs within 2 years (four crop growing seasons) depending on type of crops, reliability of RHM system and economy of scale. Improved RHM systems would reduce hydrological risks and farmers' averseness to getting credits and investing in agricultural development. Therefore, hydro-economic risks that inhibit agricultural investments in semi-arid environments can be addressed by adopting improved RHM and efficient water management systems.

Finally, improved agricultural production and sustainable livelihoods of smallholder farmers in semi-arid environments cannot be achieved through the adoption of RHM systems alone. Thus the importance of agronomical practices such as timeliness of farm operations and timing of planting dates with respect to rainfall onsets cannot be over-emphasized. For instance, early planting has been found to significantly reduce the severity of dry spells during the crop growth period. Therefore, a multi-sectoral and integrated approach is a prerequisite for achieving sustainable solutions to the many problems afflicting rural livelihoods in water scarce regions. The challenge ahead is how to disseminate the results to subsistence farmers among other stakeholders with the aim of influencing the policy and decision making process both at local and national levels.

Chapter 7

7.0 Flood Storage and River Water Abstraction [12]

7.1 Overview

The upper *Ewaso Ng'iro* basin, which starts from the central highlands of Kenya and stretch northwards transcending different climatic zones, has experienced decreasing river flows for the last two decades. The *Naro Moru* sub-basin is used to demonstrate the looming water crisis in this water scarce river basin. The decreasing flows are attributed to over-abstraction mainly for irrigating horticultural crops. The number of abstractors has increased 4 times over a period of 10 years. The quantity abstracted has also increased by 64% over the last 5 years. Ironically, the proportion of unauthorized abstractions has been increasing over the years, currently at about 80% and 95% during high and low flows respectively. This has resulted in alarming conflicts among various water users. The situation is aggravated by low irrigation efficiency (25-40%) and lack of flood flow storage facilities. The chapter analyzes over 40 years' observed river flow data and 5-year interval water abstraction monitoring records for 15 years.

The chapter assesses whether RHM systems, in particular flood storage, can reduce dry seasons' irrigation water abstractions without significantly reducing river flows to affect the sustenance of natural ecosystems downstream. The results demonstrate that flood storage can reduce abstraction and increase river flows during the dry seasons. However, financial and hydrological implications should be considered if a sustainable basin water resources management strategy is to be developed and implemented. The case study of *Naro Moru* sub-basin is representative of the situation in the other sub-basins, and hence can be taken as a pilot basin for developing integrated water resources management strategies that will foster socio-economic development with minimal negative hydrological impacts in the water stressed upper *Ewaso Ng'iro* river basin.

The water demand is continuously increasing due to population growth and irrigation development. Farmers migrating from adjacent high agricultural potential districts, due to pressure on land have resulted in land use change in the lower zones from natural vegetation to small-scale agriculture, which have led to increased water abstraction and drastically decreased river flows. There was substantial increase in small-scale farming from 1984-1992, which led to decrease in grassland and grassland mixed with trees (Roth, 1997). The main land uses in the sub-basin are given in Table 7.1. Current settlers practice a combination of rainfed and irrigated agriculture, mainly on subsistence basis growing crops such as maize, beans, cabbages and potatoes. Irrigated agriculture takes a more commercial

---

[12] *Based on:* Ngigi, S.N., H.H.G. Savenije and F.N. Gichuki. (*forthcoming*). Hydrological impacts of rainwater harvesting and management on dry seasons' irrigation water abstraction in upper *Ewaso Ng'iro* river basin, Kenya. The paper has been submitted to *Agriculture, Ecosystems and Environment*

perspective mainly targeting horticultural crops, which have ready markets—out-growers for existing large-scale export oriented companies. Farmers in the savannah zone have also been adopting RHM systems such as conservation tillage, on-farm runoff storage (water pans/ponds) and flood diversion and storage for livestock and irrigation.

Table 7.1. Types of land use and their coverage in *Naro Moru* sub-basin

Type of land use	Area (km^2)	Proportion of sub-basin (%)
Ice cover	1.25	0.72
Natural forest	35.83	20.71
Planted forest	2.24	1.29
Cropland	39.22	22.67
Rainfed	*24.22*	*8.67*
Irrigated	*15.00*	*14.00*
Grassland	94.35	54.54
Urban	0.04	0.02
Water body	0.07	0.04
Total	*173*	*100*

Source: Adapted from Niederer (2000) and NRM (2003)

Other water sources with limited exploitation are groundwater and water pans mainly in the lower zones. There are 5 boreholes (for domestic water and livestock) and 7 water pans (for domestic, livestock and irrigation) (NRM, 2003). The allocation and control of diminishing water supply within the catchment is a major challenge due to related hydrological, environmental and socio-economic implications. This calls for proper water management to ensure that this resource is used in a sustainable way for the benefit of all users. In the past, emphasis has been on using river water to meet demand but there is an urgent need to devise options to manage the increasing demand. Some options include improving water use efficiency, soil moisture conservation in rainfed agriculture, restricting water use during critical dry periods and wet season flood flow diversion and storage for use during the dry seasons. However, sustainable solutions to addressing ensuing conflicts over water among different users rely on formulation of adaptive policies and strategies. It remains to be seen if RHM can substantially reduce dry season irrigation water abstractions, which is the major cause of reduced river flows and conflicts among downstream and upstream users. It is evident that recent land use changes could be responsible for higher flood flows due to increased runoff generation in the savannah zone (NRM, 2003). Management of high flood flows through RHM would not only reduce its negative effects such as soil erosion, but also reduce water demands and abstractions during the dry seasons.

7.2 Hydrological Monitoring and Data Analysis

The *Naro Moru* sub-basin is the best documented river in the upper *Ewaso Ng'iro* basin, with a hydro-meteorological monitoring network consisting of 6 river gauging stations (A1-A6), 8 rain gauges and 4 evaporation pans (Decurtins, 1992), which have been effectively managed by the NRM project based in *Nanyuki*. The sub-basin has a long term record spanning back to 6 decades, i.e. since 1931 (Leibundgut, 1986). Fig. 7.1 shows rainfall distribution within the sub-basin. Observed river flow data show that there has been a progressive decrease in river flows and increasing water conflicts, which can be attributed to increasing water abstractions, especially for irrigation during the dry seasons. Comprehensive water abstractions monitoring was done in 1992 (Gathenya, 1993), in 1997 (Gikonyo, 1997) and in 2002 (NRM, 2003). The number of abstraction points has increased from 26 in 1990 (Gathenya *et al.*, 2000) to 76 in 1997 and to 100 in 2002 (NRM, 2003; Aeschbacher *et al.*, 2005) along the settled 30km section of *Naro Moru* river, mainly below the forest zone (see Fig. 7.2).

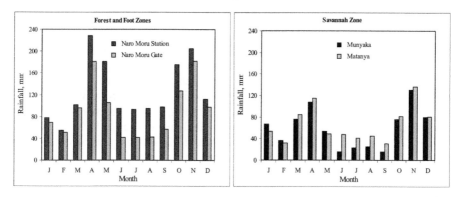

Fig. 7.1. Rainfall distribution in *Naro Moru* sub-basin at different river reaches (1980-2004)

The increase can be attributed to many small-scale farmers using portable pumps. However, community irrigation projects and large-scale horticultural farmers have also increased the quantity of water abstraction. Surprisingly, about 80% of the total abstraction volume is carried out by 5 abstractors—3 community irrigation schemes and 2 large horticultural farms (NRM, 2003). Table 7.2 presents the types of abstraction and quantity of water abstracted. The quantity of abstracted water was measured using cut-throat flumes and current meters for furrows, calibrated bucket and stop watch for gravity pipes and pumping rates, capacity and pumping duration (Gathenya *et al.*, 2000; NRM, 2003). For large abstraction points, river flow measurements were made by use of current meter upstream and downstream of the abstraction point. Amount of fuel used by pumps can also be used to estimate the amount of water abstracted.

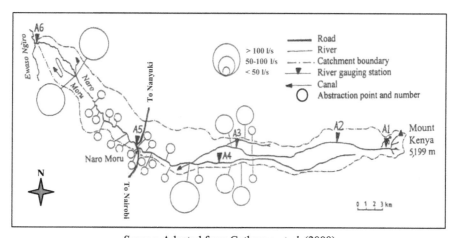

Source: Adapted from Gathenya *et al.* (2000)

Fig. 7.2. River gauging stations and some of the water abstraction points in *Naro Moru* basin

Demand based estimates were used where abstraction measurements were not possible. For example, the types of crops, cropping pattern, acreage and irrigation methods were used to estimate irrigation water demand. According to Aeschbacher *et al.* (2005), most of the abstracted water is used for irrigation (97%), while the rest is for livestock (2%) and domestic (1%) use. The area under irrigation is about 15km^2, i.e. 8% of the total catchment area. Water abstraction assessment revealed that about 62% of the dry season flows and 43% of the wet season flows are abstracted from *Naro Moru* river before its confluence with the *Ewaso Ng'iro* river. Though the river is perennial, over-abstractions (Gichuki *et al.*, 1998 and Gikonyo, 1997) leads to drying up of the lower reach (below A5) during the driest months of February and March, and under extreme conditions from July-September.

The proportions of unauthorized or illegal abstractions have been increasing over the years. Illegal abstraction here refers to either abstracting without a permit or abstracting above the permitted limits. According to Aeschbacher *et al.* (2005) about 80% is abstracted illegally during flood flows and up to 98% during low flows. Gathenya *et al.* (2000) reported 70% and 92% illegal abstractions during flood and low flows respectively. Moreover, out of a total of estimated water abstractions of 0.74 m^{3}s^{-1}, existing permits indicate that only 0.16 m^{3}s^{-1} and 0.014 m^{3}s^{-1} are allowed during flood and normal flows respectively (NRM, 2003). A similar study conducted in 1997 revealed that total river water abstraction was 0.45 m^{3}s^{-1}, and hence a 64% increase over a period of 5 years. The situation is expected to worsen as demand of horticultural crops continues to increase and more land is put under irrigation without due consideration of water availability. This means the natural environment, and the biodiversity it contains, is threatened by increasing water withdrawals (UNESCO and WMO, 2001).

Table 7.2. Types of abstraction and quantity of water abstracted in 2002

Type of abstraction	No. of abstractions	Abstracted quantity (m^{3}s^{-1})	Quantity (%)
Furrows	6	0.28	38
Gravity pipes	7	0.27	37
Fixed pumps	8	0.03	4
Portable pumps	79	0.16	21

Source: Adapted from NRM (2003) and Aeschbacher *et al.* (2005)

7.3 River Flow Trend and Duration Curves

7.3.1 Trends in river flows

The total water abstraction from *Naro Moru* river shows a significant correlation with the river discharge (correlation coefficient of 0.91, R^2=0.83, a=5) while abstraction points above the foot zone (A5) are not significantly dependent on river slows (Aeschbacher *et al.*, 2005). Fig. 7.3 (a) shows that the average river flows on the lower river reach has gradually been decreasing, while the upper reach indicates no significant decline. The specific river flows (or yields) in ls^{-1}km^{-2} give an overview of the spatial distribution of the water quantities which are available in the rive basins and allows the direct comparison between the discharges of the individual watersheds (Leibundgut, 1986) as shown in Fig. 7.3(a). River flow during the dry season is only enough for domestic and livestock water needs and for minor irrigation on a kitchen garden scale. Thus, increasing irrigation water demands can only be met if RHM systems (on-farm storage and construction of reservoirs along the river) are considered.

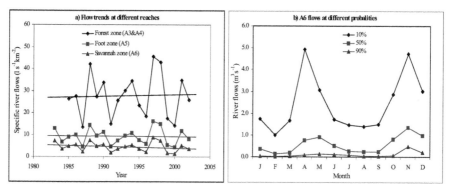

Fig. 7.3. Annual and monthly river flow characteristics of *Naro Moru* sub-basin

River flow analysis at the lower zones (foot zone and savannah), i.e. downstream from A3&A4, indicates high variability and temporal fluctuations as shown in Table 7.3. In general, Table 7.3 shows the same pattern as Fig. 7.3(a). However, Fig. 7.3(a) shows a recurrence of extreme conditions of low flows in cycles of 3-5 years, i.e. on average once in every 4 years. Fig. 7.3(b) shows that the river exhibits a bimodal pattern, with peaks corresponding with the two main rainy seasons (Fig. 7.1). During the dry season, river flows are maintained by melting glaciers and groundwater seepage (Gichuki *et al.*, 1998).

Table 7.3. Mean flow (m^3s^{-1}) at different time period and river reaches of *Naro Moru* river

River Gauging Station	Catchment area (km^2)	Period							
		1983-1985		1986-1990		1991-1995		1996-2000	
		mean	Std	mean	Std	mean	Std	mean	Std
A3&A4	48.4	1.28	1.15	1.40	1.84	1.25	1.09	1.35	1.35
A5	121.1	1.17	1.16	1.21	1.71	0.98	1.74	1.14	1.16
A6	173.0	0.92	1.04	0.92	1.45	0.66	1.56	0.75	2.01

Monthly flow data analysis (Table 7.4) shows that there is a high variability in flows with the highest flows occurring in the months of April and November, while three months—January, February and September records no flows—river dries up in these months. Though the rainfall pattern is bimodal, continental rains (July to September) sustain river flows. Table 7.3 shows a decline in river flows at A6, which can be attributed mainly to increasing water abstraction and drought cycles. Fig. 7.3(a) supports this argument since over the same period, there is no significant decline in flows at the upper river reaches.

Comparison between discharge parameters at A5, shows that although the mean flows have remained constant, Q_{50} and Q_{95} have respectively changed from 0.74 m^3s^{-1} and 0.29 m^3s^{-1} in the period 1960-1984 to 0.51 m^3s^{-1} and 0.11 m^3s^{-1} in the period 1985-2002 (Aeschbacher *et al*, 2005). Coincidently, low flows correspond with the dry season, when irrigation water demand is highest, which leads to over-abstraction to the extent that the river dries up at A6. *Naro Moru* river is reported to experience over 80% over-abstraction (Gichuki *et al.*, 1998). This in essence means even the little water that may be available is not adequate for different uses. This is made clearer by the flow duration curves at different river reaches as shown in Fig. 7.4.

Table 7.4. Monthly flow characteristics at *Naro Moru* sub-basin outlet (A6)

Month	Long term monthly flow parameters 1983-2002 (m^3s^{-1})				
	Mean	Median	Std. Deviation	Lowest	Highest
January	0.81	0.38	1.53	0.00	17.4
February	0.40	0.15	0.73	0.00	9.18
March	0.57	0.19	1.01	0.03	10.03
April	1.81	0.76	2.50	0.03	17.75
May	1.38	0.91	1.58	0.03	13.86
June	0.83	0.52	1.03	0.03	8.00
July	0.55	0.27	0.66	0.03	5.84
August	0.56	0.24	0.86	0.03	8.67
September	0.60	0.24	0.94	0.00	8.03
October	1.25	0.81	1.51	0.03	10.88
November	2.07	1.38	2.13	0.05	17.87
December	1.52	1.00	1.88	0.03	15.87

7.3.2 Flow duration curves

The flow duration curves (Fig. 7.4) show high water abstraction in the lower river reaches between the foot zone and savannah compared to the upper reaches. The difference in river flows between A6 and A3&A4 indicates more abstractions in the lower reaches. Moreover, the steeper slope of the flow duration curve at A6 indicates a higher rate of abstraction than the moderate slopes for upper zones. The high abstraction is mainly due to hydraulic advantage and settlement density. However, a proportion of water abstracted in the upper zone is used in the lower zones for irrigation and urban demand (Gichuki *et al.*, 1998). According to NRM (2003), the proportion of water abstraction as a percentage of available river flows increase from 22% in the forest zone, to 43% in the foot zone and to 61% in the savannah zone. In Table 7.5 we see that in low flow 2002 there is an increase of 28%, 50% and 77% respectively.

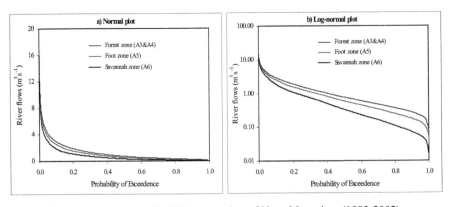

Fig. 7.4. Flow duration curves for different reaches of *Naro Moru* river (1983-2002)

7.4 Can RHM reduce Dry Season's Irrigation Water Abstraction?

Potential of RHM systems to reduce water crisis in *Ewaso Ng'iro* basin has been documented (WRAP, 1987; Thomas *et al.*, 1997; Gichuki *et al.*, 1998; Gichuki *et al.*, 1999; Gichuki, 2002). To assess the hydrological impacts of RHM (flood flow storage), the observed flows at A6 in 2002 were naturalized by adding the monthly water abstractions to the measured river flows. The naturalized flows represent the potential water available for various uses. The year 2002 was selected for this analysis partly due to its below average measured river flows—water shortage year—and partly due to existence of the most recent water abstraction records. The actual monthly water abstractions from *Naro Moru* sub-basin in 2002 were obtained from Aeschbacher *et al.* (2005). The effect of flood flows storage can be assessed from naturalized flows at the sub-basin outlet (see Table 7.5). The unaccounted for can either be attributed to losses (positive values) (mainly infiltration to groundwater, limited evaporation and undetected abstractions) or groundwater seepage (negative values). Apparently there is some groundwater recharge during wet months, which releases some flow in dry months.

Table 7.5. *Naro Moru* monthly flows, abstraction and naturalized flow in for 2002

Month	Observed river flows, abstraction and naturalized flow (m³s⁻¹)					
	A3&A4	A5	A6	Abstraction	Naturalized flow (A6)	Δflow (A6)
January	1.80	1.52	1.00	0.51	1.51	0.29
February	0.53	0.33	0.13	0.34	0.46	0.07
March	1.02	0.72	0.32	0.51	0.83	0.19
April	2.20	1.84	1.25	0.51	1.76	0.44
May	2.89	2.42	1.85	0.80	2.65	0.24
June	0.81	0.53	0.24	0.51	0.75	0.06
July	0.42	0.26	0.08	0.34	0.42	0.00
August	0.38	0.28	0.10	0.34	0.44	-0.06
September	0.30	0.18	0.06	0.34	0.40	-0.10
October	0.85	0.58	0.18	0.51	0.69	0.16
November	2.32	2.00	1.64	0.80	2.44	-0.12
December	1.51	1.34	0.92	0.51	1.43	0.08
Mean	1.25	1.00	0.65	0.50	1.15	0.10
% abstraction	28	50	77	-	-	-

The amount of flood flow to be stored depends on the dry seasons water demand. Since 97% of water abstracted is used for irrigation, the required storage capacity was determined from irrigation water requirement (m³ day⁻¹) using the method adapted from Doorenbos and Pruitt (1977) given in Eq. (7.1).

$$Q_i = \frac{K_c * E_o * A}{\eta} \qquad (7.1)$$

where; Q_i is crop water requirement (m³ day⁻¹), K_c is crop factor, E_o is reference potential evaporation (m day⁻¹), A is area (m²), and η is irrigation efficiency (%).

From the crop growing period, then the total seasonal flood storage capacity can be estimated. However, due to the high investment cost implications, it is not advisable to assume that all seasonal crop water requirements can be obtained from flood storage. For computation purposes, different proportions of crop water requirements have been considered to be from RHM—flood storage systems. The remaining proportion is obtained from river water abstractions. Since the focus is the river sub-basin, farmers who are hydrologically advantaged should be encouraged to meet their irrigation demand through flood storage. Construction of communal flood diversion and storage structures should also be considered to reduce construction cost.

The flood storage would only be considered during the months of high flows, i.e. April/May and November/December. The stored amount will be used for irrigation during the following dry seasons of 90-120 days, i.e. January-March and June-September, which are adequate for most vegetables and horticultural produce. However, irrigation water requirements can be reduced by staggering cropping pattern to ensure the crops take advantage of the wet periods. Management of cropping patterns can also ensure good market prices. By considering various options of meeting irrigation water demand, the proportion of river flows that can be allowed downstream can be determined. Increasing flood storage will reduce river water abstractions during the dry seasons, and hence increase dry season river flows. However, the amount of flood storage should not reduce high flows below certain limits—determined by the requirements of natural ecosystems and recharging of groundwater reservoirs.

Therefore, balancing the amount of flood storage and dry season abstraction is a prerequisite for effective water management. For E_o = 5mm day^{-1}, K_c = 0.8, and η = 50%, Eq. (7.1) gives an irrigation water requirement of about 1.40 m^{3}s^{-1} over the total irrigated area of 1500 ha. If this amount (i.e. no dry season water abstraction, i.e.100% flood storage) is to be stored during 3 high flow months for use during the 3 months of dry season, then each of the high flow months will store, on average, 1.40 m^{3}s^{-1}. Assuming 50% storage efficiency, this translates to 2.8 m^{3}s^{-1} per rainfall season. Thus for 3 months on-farm storage of 14,500 (2.8*86400*30*3/1500) m^3 ha^{-1} if losses are controlled is required or, 7,250 m^3 ha^{-1}. The storage capacity can be reduced further to 4,830 m^3 ha^{-1} if irrigation efficiency is improved from 50% to 75% by adopting drip irrigation technology. Table 7.6 shows levels of flood storage at different irrigation and storage efficiencies. Smaller storage structures would suffice if supplementary, instead of full irrigation is practised. Less flood storage would mean less investment costs, and hence a compromise can be attained depending on both financial and hydrological implications.

Table 7.6. Flood storage (m^3 s^{-1}) at different irrigation and storage efficiencies

Irrigation efficiency (%)	Flood storage efficiency (%)			
	25	50	75	100
25	11.20	5.60	4.20	2.80
50	5.60	2.80	2.10	1.40
75	4.20	2.10	1.40	1.05
100	2.80	1.40	1.05	0.70

A comparison between average abstractions ((0.5 m^{3}s^{-1}) from Aeschbascher *et al.*, 2005) and computed net irrigation water demand (0.7 m^{3}s^{-1}) shows that only 71% of the irrigable land could be irrigated even at 100% irrigation efficiency in 2002. This could either be due to low flows or overestimation of actual irrigated area. Otherwise the crops are under severe moisture stress, which could lead to low production, despite high cumulative abstraction percentage. Due to inadequate water, farmers are forced to either reduce their irrigated area or reduce amount of irrigation water. This could affect production and investments. Therefore, given that farmers abstracted all the available flow, the actual abstracted amounts (0.5 m^{3}s^{-1}) were used for flood storage analysis in Table 7.7 and Fig. 7.5.

Flood storage analysis shows that it is possible to maintain average downstream river flows and increase irrigation water supply—ensuring adequate downstream water flows while meeting the irrigation needs of upstream farmers through flood storage. However, it is not possible to store adequate flood flows to meet the required irrigation water demand during low flow years. It appears already the irrigation water demand is beyond what the river flows can supply. Fig. 7.5 presents the balance between two opposing strategies in flood storage; (a) priority given to storage demand, i.e. water released downstream only after storage requirements are met, and (b) priority given to DFR, i.e. proportioning river flows to ensure there is adequate flow for downstream users in each month. Fig. 7.5 shows the river flow patterns for naturalized flows, observed flows and simulated flows for different proportion (50% and 75%) of flood storage at A6 based on the two strategies.

Table 7.7. Impacts of flood storage on *Naro Moru* river flow (m^3s^{-1})

Month	Naturalized flow	Measured flow	Scenario (a)		Scenario (b)	
			50%	75%	50%	75%
January	1.51	1.00	1.01	1.26	0.68	1.09
February	0.46	0.13	0.00	0.21	0.21	0.33
March	0.83	0.32	0.33	0.58	0.38	0.60
April	1.76	1.25	0.34	0.17	1.12	0.80
May	2.65	1.85	2.58	2.00	1.68	1.19
June	0.75	0.24	0.45	0.6	0.34	0.54
July	0.42	0.08	0.12	0.27	0.19	0.30
August	0.44	0.10	0.14	0.29	0.20	0.32
September	0.40	0.06	0.10	0.25	0.18	0.29
October	0.69	0.18	0.39	0.54	0.31	0.50
November	2.44	1.64	0.94	0.24	1.55	1.11
December	1.43	0.92	1.01	1.39	0.91	0.65
Mean	1.15	0.65	0.65	0.65	0.65	0.65

The management rules under the two strategies are: (a) during the wet season, flood flow is stored for use in the dry season, with the percentage storage based on a set amount to be supplied from storage e.g. 50% flood storage means adequate flood flow will be stored to meet 50% of irrigation water supply during the dry season; (b) during the dry season, no flow is stored and the amount of naturalized flow released downstream is a function of flood storage level e.g. 75% flood storage means that only the 25% of irrigation water demand will be abstracted from river flows.

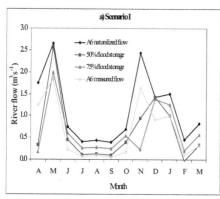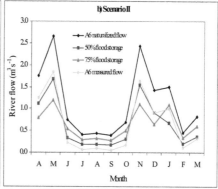

Fig. 7.5. The effects of flood storage on *Naro Moru* river flow at A6 in 2002

Fig. 7.5 presents the results of flood storage analysis which are detailed in Annexes 4-6. Two different scenarios for filling reservoir are distinguished in Figs. 7.5a and 7.5b. In scenario I, the storage is allowed to fill at first opportunity, while in scenario II proportionate filling is allowed depending on monthly flows based on long term flow patterns. As a result, the downstream hydrographs in Fig. 7.5a show less of a shock and do not suffer from the very low flows during the months of the two rainy seasons (i.e. April and November). However, it may not be possible to store all the initial flood flows because between A6 and the reservoir, which hydraulically can be located above A5, there will be substantial runoff contributing to flow downstream. The lower zones contribute substantial surface runoff during the rainy season; hence the situation in Fig. 7.5a will allow flows equivalent of the difference of naturalized flows at A5 and A6. Similarly, scenario II would also allow for progressively increasing flow from downstream catchments.

This reality is hidden in monthly flows, but daily flows show the contribution of the lower catchments, i.e. there are higher flows at A6 than A5 and A4&A3 for high rainfall events (see Fig. 7.6). The monthly averages may not show clearly what happens during high storms, especially in the savannah zone. Fig. 7.6 shows daily river flows superimposed on daily rainfall of *Matanya*, which is in the savannah zone. The flood storage analysis also assumes that one big reservoir will be constructed to regulate the river flows, but the storage system will consist of a series of individual and communal storage reservoirs spread along the river course. Hence scenario II is more in agreement with reality than scenario I.

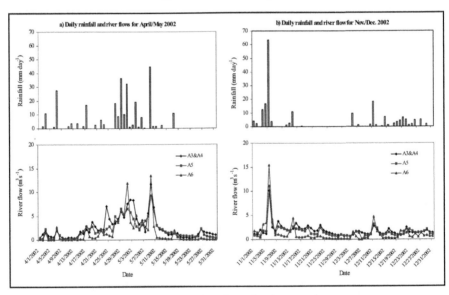

Fig. 7.6. Daily rainfall for *Matanya* and river flows for the two rainy seasons

Nevertheless, both scenarios show similar redistribution of river flows, which allows more downstream flows during the dry season. Therefore, flood storage may reduce dry season water abstractions, without significantly reducing flood flows which are important in sustaining natural ecosystems and groundwater recharge downstream. For example, if only half of the crop water requirement is to be supplied from RHM system, i.e. 50% flood storage, then 50% of dry season abstraction will be released to downstream users. Flood storage does not affect the overall average flows, but distribute it without reducing the cumulative flows. Thus the gains in reducing dry season's abstractions would supersede the impacts of reduced flood flows. The amount of flood storage can be reduced by improving storage efficiency, irrigation efficiency (currently at 25-40%), allowing minimal dry season abstractions and in-situ soil moisture conservation, which will reduce irrigation water requirement being met by stored flood and unproductive water losses. Since 2002 was a low flow year (Fig. 7.3), the positive impact of flood storage will be less for above normal flow years. Other benefits of flood storage are groundwater recharge, reduced cost of pumping and improved crop management. Some of the losses, especially seepage, may eventually contribute to groundwater recharge. The cost of pumping can also be reduced by locating flood storage structures where water can be delivered by gravity to the farms.

Technically, besides the financial constraint related to the cost of storage facilities, flood harvesting for dry season irrigation is feasible. The financial and hydrological implications should however, be considered in the formulation and implementation of a sustainable sub-basin water resources management strategy. The option of flood storage has already been adopted by some large-scale horticultural farms; for example, *Homegrown*, which has more than 300 ha of irrigated horticulture in *Timau* for export, stores up to 742,000 m^3 of flood flows in four earthdams to sustain its irrigation water demand (Ngigi, 2003a).

7.5 Implications on Upper *Ewaso Ng'iro* Basin

The water crisis experienced in *Naro Moru* sub-basin is typical of what is happening in the other sub-basins. The excessive water abstractions for irrigation upstream have resulted in diminishing river flows, which are finally reflected by discharge of the entire upper *Ewaso Ng'iro* basin. Recent studies on three sub-catchments (*Burguret*, *Likii* and *Timau*) revealed similar hydrological trends (Liniger *et al.*, 2005). The overall hydrological implication is reduced river flows, increased conflicts over water among stakeholders and negative impacts on natural ecosystems that thrive on sustained flows and periodical flooding. Fig. 7.7 shows a decreasing trend of *Ewaso Ng'iro* river at *Archer's Post* over the last 4 decades.

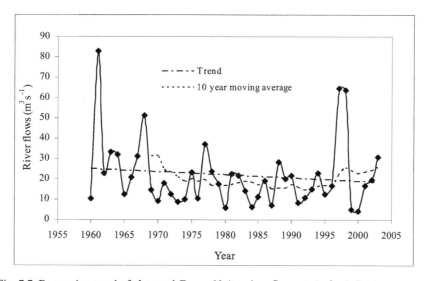

Fig. 7.7. Decreasing trend of observed *Ewaso Ng'iro* river flows at *Archer's Post*

The 10-year running average indicates that the flow has been decreasing since 1970. The months with the lowest flows correspond with those at the sub-basins; hence activities at the upper reaches affect the total river basin discharge. River flows at *Archer's Post* are lowest in February of which the mean has decreased from 9.0 m^3s^{-1} in the 1960s, to 4.6 m^3s^{-1} in the 1970s, to 1.3 m^3s^{-1} in the 1980s, and to 1.0 m^3s^{-1} in the 1990s (Liniger, 1995). A section of *Ewaso Ng'iro* river dried up completely in 1984, 1986, 1991, 1997 and 2000 (Gichuki *et al.*, 1998; Gichuki, 2002). The trend line (Fig. 7.7) shows that the annual mean has decreased from 25 m^3s^{-1} in 1960 to 18 m^3s^{-1} in 2002, which is a 28% flow reduction.

The cause of the decreasing river flows is mainly induced by human activities around the slopes of Mt. Kenya and *Nyandarua* Ranges since there is no corresponding decline in rainfall amounts over the same period (Gichuki, 2002). The trend is alarming and water scarcity related conflicts among different groups are increasingly becoming common. Although not all reported conflicts are related to water scarcity, the majority is water related. Water scarcity and associated conflicts have been compounded by lack of good understanding of the nature and extent of water shortage, failure to take action required to address water scarcity issues, inequity in resources access, poverty, lack of alternative sources of livelihoods, high cost of water resources development and technologies that use water efficiently, political interference, stakeholders' diverse perceptions on water availability, their entitlements and needs of other water users (Gichuki, 2002), and inadequate policy, legal and institutional framework. However, our interest here is how we can manage the limited water resources for socio-economic development and ecological needs. We believe that flood storage and management can be one of the sustainable solutions if supported by responsive policies and institutions that will adopt integrated water resources management principles and embrace direct and indirect actors and stakeholders.

7.6 Conclusions

The decreasing river flow in upper *Ewaso Ng'iro* river basin is alarming and urgent attention is required to reverse the trend. The manifestation at the sub-basins is reflected by flow characteristics at the river basin outlet. The case of *Naro Moru* sub-basin has been used to demonstrate this scenario and represent all the related constraints, challenges and hydrological and socio-economic impacts of management in a water scarce basin. *Naro Moru*, like other sub-basins of the *Ewaso Ng'iro* river basin has experienced heavy immigration of settlers from the adjacent high potential districts since the early 1970s. This explains the decreasing trends of river flows (about 30% reduction in 40 years at *Archer's Post*) against an ever increasing water demand and pressure on other water resources. This has led to high proportions of unauthorized water abstractions. For instance, out of the 100 current water abstractors, only 25 have water permits out of whom, only 12 have been authenticated by the ministry of water and irrigation.

The following are some of the reasons that have been attributed to high unauthorized abstractions: inadequate and ineffective water abstraction monitoring systems; high financial returns from irrigated agriculture; low fines for illegal abstractions; lack of floodwater storage facilities; and low water use efficiency for irrigation (25-40%) for smallholder irrigation schemes. Improving agricultural water management can increase food production by rising water-use efficiency, both for irrigated and rainfed agriculture. Despite the high water abstractions, river flows cannot meet irrigation water demand, even at 100% irrigation efficiency. It seems either part of the irrigable land is irrigated or crops are under severe moisture stress. The available water can only adequately irrigate about 40% of irrigable land at 50% irrigation efficiency. More land can be put under irrigation by either improving irrigation efficiency and/or incorporating flood storage.

While there is no doubt establishment of water users' associations suggested by past studies would improve the management of water resources in *Ewaso Ng'iro* basin, without developing means of reducing water demand and increasing flood water storage such noble efforts may be fruitless. A multi-sectoral and integrated approach would be a prerequisite towards formulating sustainable water management strategies. However, without providing alternative sources of water and effective demand management mechanisms that will reduce over-abstraction this may not be easily achievable.

The results show that RHM can provide an alternative water source, which may reduce the demand on river flows and over-abstraction during dry seasons. RHM systems can be either small to medium on-farm storage structures such as farm ponds or large community off-stream dams[13]. This means that excess runoff and flood flows would be stored for use during low flows. Besides reducing dry season's water abstraction, flood storage will make water available throughout the year—even distribution of water. This will address conflicts between irrigation and other essential water uses. Irrigation permits should stipulate that abstractions only take place during flood flows and that water stored for use when required. One way of ensuring compliance would be to make it mandatory for permit applicants to first construct a 90-day storage as stipulated by the Water Act and subsequent legislation (GoK, 1972; GoK, 2002b).

RHM also have another positive environmental impact. The high runoff generated leads to land degradation and if tapped would reduce soil erosion and siltation of water bodies. Associated soil nutrients losses will also be reduced. However, this strategy would require heavy investments, which may be a constraint to the resource-poor smallholder farmers. Here is where the sub-basin water resources management institutions, such as water users' association may need to be strengthened to implement such a strategy. The government can facilitate the implementation of catchment based integrated water resources management action plans, similar to the success stories of India (Agarwal and Narain, 1997; Agarwal *et al.*, 2001). The government facilitation can be in terms of responsive policies and incentives to private and public sector investors in water resources development and management, and provide credit to resource-poor farmers to construct off-stream and on-farm rainwater storage systems. Although according to UNESCO (2005), subsidies may lead to inefficient water use, it should be considered as an incentive for constructing flood storage structures. Therefore, one of the sustainable solutions for addressing diminishing water resources and increasing demand in *Ewaso Ng'iro* river basin is within our reach.

In general, the positive hydrological impacts of RHM in upper *Ewaso Ng'iro* river basin outweigh the associated negative impacts: reduced river flows during high floods and high construction costs. The amount of river flows reduction would however be insignificant compared to the ensuing over-abstraction during the dry

---

[13] Note that on-stream storage structures, which have been the conventional storage systems, are not desirable since they also retain river flows during the dry season and thus lead to more water management challenges and conflicts. Nevertheless, individual or family farm ponds should be encouraged to enhance proper management and maintenance.

periods if no storage is provided. The reduction of flood flows may be inconsequential compared to the reduction of dry season flow. However, it is important to ensure that socio-economic activities upstream sustain adequate river flows for downstream water uses. This may provide some insights to one of the research questions "how to increase irrigated agricultural production without significantly reducing river flows and affect hydro-ecological sustainability downstream?" Thus the challenge is to understand how water is used in different parts of a river basin in order to improve overall river basin water resources management. Therefore, to address this challenge, detailed hydrological monitoring and modelling would be necessary to provide more conclusive results.

Chapter 8

8.0 Conclusions and Recommendations

The study aimed to identify and evaluate land and water management systems, in particular RHM systems, for upgrading rainfed agriculture and also assess their hydrological impacts on water resources management and socio-economic development of upper *Ewaso Ng'iro* river basin in Kenya. A comparative review and analysis of RHM systems in GHA was done through six case studies conducted in Ethiopia, Kenya, Tanzania and Uganda. One of the case studies was conducted in semi-arid *Laikipia* district, which covers 50% of upper *Ewaso Ng'iro* river basin. RHM systems were categorized into three based on rainfall portioning (i.e. in-situ and runoff-based systems), water storage and application (i.e., direct application and storage systems) and size of catchments (i.e. micro-, small- and macro-catchment systems). However, detailed agro-hydrological analysis was carried out on small on-farm storage systems and in-situ systems (conservation tillage), which are common in *Laikipia* and have high potential for smallholder farmers. Agro-hydrological analysis was based on a conceptual framework that was developed to assess hydrological impacts of up-scaling RHM systems in a river basin.

Agro-hydrological analysis of on-farm storage systems revealed that the sizes of farm ponds ranged 30-100m^3 and their catchments vary from 0.3-2.0 ha. The catchments were found to generate adequate runoff to meet SIR for a 300-600m^2 kitchen garden, if water losses (seepage and evaporation), which accounted for 30-50% of seasonal water storage were controlled. Evaporation losses ranged from 0.1-0.3m^3 day^{-1} and seepage losses from 0.03-0.4m^3 day^{-1} on clay soils and more than 2.0m^3 day^{-1} on sandy soils. To reduce seepage losses, lining the pond surface with ultra-violet resistant plastic sheet was found to be effective, while water management was improved by incorporating low-head drip irrigation system. Agro-hydrological evaluation of in-situ RHM systems focused on conservation tillage and entailed estimation of the amount of runoff retained on cropland and incremental grain yields. Field data based on runoff and grain yield measurements were compared with empirical approach—crop water response to water. The cost of equipment and time consumed in field monitoring necessitated the development a simple empirical method, which compared well with field results.

From the positive results, in terms of yield improvements (increase and stabilization), it was found necessary to carry out hydro-economic analysis to understand why the rate of adoption was low. Hydro-economic evaluation entailed assessment of hydrological hazards and economic factors that affect investment options of risk-averse farmers in semi-arid agro-systems. Hydrological risks are related to low and poorly distribution rainfall, which leads occurrence of intra- and off-seasonal dry spells. Low rainfall reliability and occurrences of dry spells are responsible for persistent crop failure. There is a 60% and 50-80% probability of occurrence of below average rainfall and agricultural droughts respectively. Occurrence of dry spells of more than 10 days is as high as 80% during critical crop growth stage. The rainfall duration is also shorter than most crop growth period. On-farm storage RHM system can adequately address this critical water deficit by storing runoff for supplemental irrigation using drip irrigation. Benefit-cost analysis revealed that a farmer can be able to recover full investment cost with 4-8 seasons, while making substantial net revenue during the repayment period. However, despite economic viability and potential for improving agricultural productivity and livelihoods, their adoption is constrained by poor performance and socio-economic status of the farmers. Hydrological performance can be improved through adequate design, water losses control and improved water management, while provision of credit facilities should address the financial constraints. However, optimal benefits can only be achieved by incorporating appropriate agronomic practices such as land and crop husbandry. For instance, early planting was found to minimize the severity of dry spells by 15-25% during the critical crop growth stage. Thus, field evaluation and benefit-cost analysis show that hypothesis 1 can be accepted.

What is the limit of up-scaling RHM systems in a river basin? Assessment on the proportion of runoff retained by smallholder RHM systems revealed that there is no significant reduction in river flow associated with adoption of these field scale systems. RHM systems retain 20-30% of runoff during high flows and their 10-15% adoption level is also low. Moreover, comparing the less than 20% area under cultivation in the river basin, the overall reduction in river flow is negligible. Hence the limit of up-scaling smallholder RHM systems may not be an issue in developing IWRM strategies for water-scarce upper *Ewaso Ng'iro* river basin. What is the issue is irrigated agriculture, which although accounts for 5-10% of cropland, uses more than 95% of river flows. Irrigation water abstraction has progressively reduced river flows at *Archer's Post* by about 30% from 1960 to 2002. The cause of flow reduction is mainly water abstractions for irrigation upstream since there is no corresponding decline in rainfall amounts over the same period. Therefore, hypothesis 2 can be rejected for small-scale RHM systems for supplemental irrigation, but accepted for large-scale RHM systems for commercial irrigation.

The *Naro Moru* sub-basin, which is one of the affected by over-abstraction was used to assess whether RHM can reduce dry season water abstraction. The assessment was based on 5 year interval water abstraction data collected since 1992. The RHM system that can reduce water abstraction is flood storage and utilization to meet some of the crop water requirement. Already some of the high water consuming horticultural farms have adopted this technology as a means of increased their production. The analysis revealed that RHM can reduce dry season river abstractions by 50-100%, depending on flood flow conditions, without drastically affecting downstream flow requirements. For example, the flow for 2002, which was a low flow year, indicated that up to 75% of high flow can be stored for use during the dry season. Flood storage ensures uniform river flow distribution throughout the year. However, it has a cost implication that should be considered.

The challenge is to ensure equity allocation of water among all water users while sustaining adequate flow to sustain natural ecosystems. A policy issues that need to be addressed is how to increase irrigated agricultural production without reducing river flows to the limit of affecting downstream water users and hydro-ecological sustainability. Hence the need to understand how water is used at different parts of a river basin in order to formulate and implement sustainable water resources management strategies. The concern at the moment should not be on smallholder RHM systems that have negligible hydrological impacts, but flood storage that can reduce dry season water abstractions and affect flood flows for natural ecosystems and downstream water users. Therefore, detailed hydrological modelling would be necessary to assess the long term impacts of flood storage, among other RHM systems, on the overall water resources management strategies.

Samenvatting

Het is moeilijk om de noodzaak voor verhoogde voedselproductie in de semi-aride gebieden van sub-Sahara Africa (SSA) te overdrijven. De landbouwproductie is over het algemeen laag (1 ton/ha), wat overeenkomt met de armoedegrens van 1 US$/dag. De situatie wordt nog verslechterd door de algemene waterschaarste, waarbij vele landen in SSA onder de limiet van 1000 m³/cap/yr verkeren. Hoewel de situatie er wanhopig uitziet, is nog niet alle hoop verloren. Een van de meestbelovende oplossingen moet gezocht worden in het verbeteren van de door regen gevoede landbouw middels het opzetten van systemen voor regenvalopvang en beheer (zgn. RHM systemen), waarmee de waterbeschikbaar voor landbouwproductie aanzienlijk kan toenemen. In deze studie worden veranderingen in landgebruik geanalyseerd, en in het bijzonder de adoptatie van RHM systemen, met het oog op het verbeteren van de door regen gevoede landbouw voor verhoogde voedselproductie, alsmede de hydrologische effecten die dit heeft op het waterbeheer van stroomgebieden. RHM systemen zijn veelzijdig en bestrijken zowel plaatselijke methoden om bodemvocht vast te houden als grotere systemen om oppervlakteafvoer op te vangen en vervolgens over het land te verspreiden.

Het doel van de studie is informatie te verschaffen die nodig is om integraal waterbeleid te maken voor zowel de socio-economische ontwikkeling als het onderhouden van een ecologische balans in stroomgebieden. De algemene doelstelling is om de hydrologische effecten van landgebruikveranderingen op het waterbeheer en de sociaal-economische ontwikkeling van het *Upper Ewaso Ng'iro* stroomgebied in Kenya te analyseren. Dit is gedaan middels een veldonderzoek van levensvatbare RHM systemen, haalbaarheids analyse van kleine reservoirs en plaatselijke RHM systemen, en het effect van reservoirs op de onttrekkingen in het droge seizoen. Een conceptueel kader is ontwikkeld als een instrument om de hydrologische effecten van veranderingen in het landgebruik in het stroomgebied van de *Upper Ewaso Ng'iro* te analyseren. De studie is uitgevoerd in semi-aride en matig-aride gebieden waar recente veranderingen in het landgebruik hebben geleid tot een afname in de rivierafvoer en vervolgens conflicten tussen watergebruikers.

De resultaten van de veldstudie laten zien dat er een aantal levensvatbare RHM systemen zijn die de landbouwproductie in een semi-aride omgeving kunnen verbeteren. Echter, hun effectiviteit is beperkt door hoge waterverliezen (30-50%), onvoldoende bergingscapaciteit (25-50% betrouwbaarheid by reservoirs van 30-50 m³), slecht waterbeheer, een hoge frequentie van droge periodes van 10-15 dagen in het groeiseizoen, en het hoge risico dat boeren lopen om te investeren in nieuwe RHM systemen. Toch heeft de studie laten zien dat zowel kleine reservoirs voor supplementaire irrigatie als plaatselijke alternatieve technieken om bodemvocht vast te houden (zgn conservation tillage) economisch aantrekkelijk zijn voor de kleine boer. Een economische analyse laat zien dat een boer zijn investering in 2-3 jaar terugverdient.

De analyse laat zien dat RHM systemen 20-30% van de oppervlakteafvoer kunnen tegenhouden, wat gezien hun geringe verbreidheid een verwaarloosbare invloed heeft op de totale afvoer van het stroomgebied. Het landbouwoppervlak in het hele stroomgebied is minder dan 20% en de graad van toepassing van RHM systemen is 10-15%. Dientengevolge zal het vergroten van het gebruik van RHM systemen geen substantieel effect hebben op het afvoerregime van de rivier. Echter het opslaan van water tijdens de natte tijd kan de onttrekkingen in de droge tijd verminderen, zonder dat dit negatieve invloed heeft op de eco-hydrologische functies benedenstrooms. Hoewel het geirrigeerde oppervlak klein (< 5%) is, is het verantwoordelijk vor 95% van de wateronttrekkingen. De lage rivierafvoer heeft de grootschalige tuinbouwers gedwongen om reservoirs te bouwen waarin tijdens hoogwater water wordt opgeslagen. Er is dus een noodzxaak om een duurzame strategie te ontwerpen om een eerlijke verdeling van het water te verzekeren. Hydrologisch modelleren is hiervoor essentieel.

References

African Development Fund (ADF). 2005. *Ewaso Ng'iro* North Natural Resources Conservation Project (ENNNRCP). Appraisal Report. Agricultural and Rural Development Department, North East and South Region, African Development Fund, ONAR, Feb. 2005

Aeschbacher, J., H.P. Liniger and R. Weingartner. 2005. River water shortage in a highland-lowland system: A case study of the impacts of water abstraction in the Mount Kenya region. *Mountain Research and Development, 25(2): 155-162*

Agarwal, A., S. Narain and I. Khurana. 2001. Making water everybody's business: Practice and policy of water harvesting. Centre for Science and Environment (CSE), India.

Agarwal, A. and S. Narain. 1997. Dying wisdom: The rise, fall and potential of India's traditional water harvesting systems. Centre for Science and Environment, Thomson Press Ltd., Faridada.

Allen, R.G., L.S. Pereira, D. Raes and M. Smith. 1998. Crop evapotranspiration: Guidelines for computing crop water requirements. FAO Irrigation and Drainage Paper 56. FAO, Rome, Italy.

Barron, J. 2004. Dry spell mitigation to upgrade semi-arid rainfed agriculture: Water harvesting and soil nutrient management for smallholder maize cultivation in *Machakos*, Kenya. Doctoral thesis in natural resources management. Department of Systems Ecology, Stockholm University.

Barron, J., J. Rockström, F.N. Gichuki and N. Hatibu. 2003. Dry spell analysis and maize yields for two semi-arid locations in East Africa. *Agricultural and Forest Meteorology 117:23-37*

Barron, J., J. Rockström and F. Gichuki. 1999. Rain water management for dry spell mitigation in semi-arid Kenya. *E. Afr. Agric. For. J., 65(1): 57-69*

Ben-Asher, J. and P.R. Berliner. 1994. Runoff irrigation. *In:* Tanji, K.K. and B. Yaron (*eds.*). 1994. Management of Water Use for Agriculture, Management of Water Use in Agriculture. *Advanced Series in Agricultural Sciences No. 22: Springer Verlag,* Berlin, Germany, pp. *26-154.*

Benites, J., E. Chuma, R. Fowler, J. Kienzle, K. Molapong, J. Manu, I. Nyagumbo, K. Steiner, and R. van Veenhuizen (eds.). 1998. Conservation tillage for sustainable agriculture. Proc. international workshop, Harare, 22-27 June, 1998. Part 1 (Workshop Report). GTZ, Eschborn, Germany. p 59

Berger, P. 1989. Rainfall and agro-climatology of the *Laikipia* plateau, Kenya. *African Studies Series A7.* Institute of Geography, University of *Berne,* Switzerland.

Critchley, W. 1999. Promoting farmer innovation: Harnessing local environmental knowledge in East Africa. Workshop Report No. 2. RELMA and UNDP, Nairobi, Kenya.

Critchley, W. 1987. Some lessons from water harvesting in Sub-Saharan Africa. Report from a workshop held in Baringo, Kenya on 13-17 October, 1986. The World Bank Eastern and Sothern Africa Projects Department. p 58

Critchley, W. and K. Siegert. 1991. Water harvesting: A manual for the design and construction of water harvesting schemes for plant production. FAO, AGL/MISC/17/91. *http://www.fao.org/docrep/U3160E00.htm*

Cullis, A. and A. Pacey. 1992. A development dialogue: Rainwater harvesting in Turkana. Intermediate Technology, London, UK. p. 126

Dastane, N.G. 1974. Effective rainfall in irrigated agriculture. FAO Irrigation and Drainage Paper No. 25. FAO, Rome, Italy.

Decurtins, S. 1992. Hydro-geographical investigations in the Mt. Kenya sub-catchment of *Ewaso Ng'iro* river. African Studies Series, A6. *Geographica Bernensia*. University of Berne, Switzerland.

Desaules, A. 1986. The soils of Mt. Kenya semi-arid northwestern footzone and their agricultural suitability. Ph.D Thesis, University of Berne, Switzerland

Doorenbos, J. and A.H. Kassam. 1979. Yield response to water. FAO Irrigation and Drainage Paper No. 33. Rome, Italy.

Doorenbos, J. and W. Pruitt. 1977. Crop water requirements. FAO Irrigation and Drainage Paper No. 24. FAO, Rome, Italy.

Droogers, P. and G.W. Kite. 2001. Estimating productivity of water at different scales using simulation modelling. IWMI Research Report No. 53. International Water Management Institute (IWMI), Colombo, Sri Lanka.

Duckham, A.N. and G.B. Masefield. 1985. Farming Systems of the World. Praeger Publishers, New York, USA.

Evanari, M., L. Shanan and N.H. Tadmor. 1971. The Negev, the challenge of a desert. Harvard University press, Cambridge, Massachusetts, USA.

FAO. 2003. Solving water conflicts in the Mount Kenya region. CD-ROM: From the International year of mountains to the international year of freshwater. Centre for Development and Environment (CDE) and the Food and Agriculture Organization (FAO). FAO, Rome, Italy

FAO. 2002. Agriculture: Towards 2015/2030. Technical interim report, Economic and Social Department *http//www.fao.org/es/esd/at2015/toc-e.ttm*, FAO, Rome, Italy.

FAO. 1986. African agriculture: The next 25 years. Food and Agriculture Organization (FAO) of the United Nations, FAO, Rome, Italy

FAO. 1978. Report on the agro-ecological zones project: Methodology and results for Africa. World Soil Resources Report No. 48, Vol. 1. FAO, Rome, Italy.

FAO and WMO. *Undated.* Applications of climatic data for effective irrigation planning and management. Training manual on roving seminar organized by the Food and Agriculture Organization (FAO) and the World Meteorological Organization (WMO). FAO, Rome, Italy.

Fentaw, B., E. Alamerew and S. Ali. 2002. Traditional rainwater harvesting systems for food production: The case of Kobo Wereda, Northern Ethiopia. GHARP case study report. Greater Horn of Africa Rainwater Partnership (GHARP), Kenya Rainwater Association, Nairobi, Kenya.

Flug, M. 1981. Production of annual crops on micro-catchments. *In:* Dutt, R.G., F.C. Hutchinson and A.M. Garduno (*eds.*). 1981. Rainfall Collection for Agriculture in ASAL. Workshop Proceedings, Commonwealth Agricultural Bureaux, UK: 39-42.

Flurry, M. 1987. Rainfed agriculture in the central division *Laikipia* district, Kenya: Suitability, constraints and potential for providing food. *African Studies Series A6*. Institute of Geography, University of *Berne*, Switzerland.

Fox, P. and J. Rockström. 2000. Water harvesting for supplemental irrigation of cereal crops to overcome intra-seasonal dry spells in the Sahel. *Phys. Chem. Earth (B), 25(3): 289-296*

Gathenya, J.M., H.P. Liniger and F.N. Gichuki. 2000. Problems of water resources management for a basin west of Mt. Kenya: Challenge to water resource planners. *In:* Gichuki, F.N., D.N. Mungai, C.K. Gachene and D.B. Thomas (eds.).2000. Land and water management in Kenya: Towards sustainable land use. Dept. of Agricultural Engineering, University of Nairobi, Nairobi, Kenya. *pp 175-180.*

Gathenya, J.M. 1992. Water balance of sections of the *Naro Moru* river. Unpublished M.Sc thesis, Department of Agricultural Engineering, University of Nairobi

Gichuki, F.N. 2002. Water scarcity and conflicts: A case study of the Upper *Ewaso Ng'iro* North Basin. *In:* Blank, H.G., C.M. Mutero and H. Murray-Rust (*Eds.*). 2002. The changing face of irrigation in Kenya: Opportunities for anticipating change in Eastern and Southern Africa. IWMI, Colombo, Sri Lanka. *pp 113-134*

Gichuki, F.N., P.G. Gichuki and R.K. Muni. 1999. Characterization of the flow regime of *Ewaso Ng'iro* North River. *East African Agricultural and Forestry, 65(1): 70-78*

Gichuki F.N., H.P. Liniger, L. Macmillan, L. Shwilch, and J.K. Gikonyo. 1998. Scarce water: Exploring resource availability, use and improved management. *Eastern and Southern Africa Geographical, Vol. 8, Special Number.*

Gichuki, F.N., S.N. Ngigi and G.A. Mukolwe. 1997. Water management by crisis: The case of *Matanya* small-scale irrigation scheme, *Laikipia* district, Kenya. Africa Centre for Technology Studies (ACTS) Research Report. ACTS, Nairobi, Kenya.

Gathuma, M.N. 2000. Design of a rainwater harvesting system for low head drip irrigation. Unpublished final year B.Sc Report. Department of Agricultural Engineering, University of Nairobi, Nairobi, Kenya.

Gikonyo, J.K. 1997. River water abstraction monitoring for the *Ewaso Ng'iro* river basin, Kenya. Paper presented at the SPPE workshop, Madagascar, 2-6 June 1997.

Giraud, F., S. Lanini, J.D. Rinaudo, V. Petit and N. Courtois. 2002. An innovative modelling concept for IWRM linking hydrological functioning and social behaviour: The *Hérault* catchment case study, south of France. Proceedings of the International Environmental Modelling and Software Society (iEMSs), 24-27 June 2002, *Lugano*, Switzerland. *Vol. 1: pp 126-131*

Githinji, J.M. 1999. Personal communication. He was the Project Manager of Joint Relief and Rehabilitation Servives (JRRS), a local NGO operating in North Eastern Kenya.

Gittinger, J.P. 1972. Economic analysis of agricultural projects. The Economic Development Institute, International Bank for Reconstruction and Development, The World Bank. The John Hopkins University Press, Baltimore, USA.

Gould, J. and E.N. Petersen. 1991. Rainwater catchment systems for domestic supply: Design, construction and implementation. IT Publications Ltd., London, UK.

Government of Kenya (GoK). 2002a. Effective management for sustainable economic growth and poverty reduction. 9th National Development Plan 2002–2008. Government of Kenya (GoK). Government Printers, Nairobi, Kenya.

Government of Kenya (GoK). 2002b. The Water Act, Kenya Gazette Supplement No. 107 (Acts No. 9). Government Printers, Nairobi, Kenya.

Government of Kenya (GoK). 1972. The Water Act, Chapter 372, Laws of Kenya. Government Printers, Nairobi, Kenya.

Government of Kenya (GoK). 1999. National population census statistics of 1999. Government of Kenya (GoK), Government Printers, Nairobi.

Hai, T.M., A.A.H. Khan, C.K. Ong, J. Rockström and D.M. Mungai. 2004. Design, calibration and field testing of a low cost, robust pipe sampler for plot scale runoff and soil erosion assessment. Paper submitted to *Trans. ASAE.*

Hai, M. 1998. Water harvesting: An illustrative manual for development of micro-catchment techniques for crop production in dry areas. RELMA Technical Handbook no. 16. Signal Press Ltd., Nairobi, Kenya.

Hatibu, N., H.F. Mahoo and G.J. Kajiru. 2000. The role of RWH in agriculture and natural resources management: from mitigation droughts to preventing floods. In: Hatibu, N. and H.F. Mahoo (eds.). 2000. Rainwater harvesting for natural resources management: A planning guide for Tanzania. Technical Handbook No. 22. RELMA, Nairobi. 58-83

Hajkowicz, S., T. Hutton, J. McColl, W. Meyer and M. Young. 2003. Exploring future landscapes: A conceptual framework for planned change. *Land and Water Australia*. Canberra, Australia. *p 57*

Helweg, O.J. and P.N. Sharma. 1983. Optimum design of small reservoirs. *Water Resources Research. Vol. 19 (4): 881-885.*

Hoekstra, A.Y. 1998. Perspectives on water: An integrated model-based exploration of the future. International Books, Utrecht, the Netherlands. *p 356*

Huber, M. and C.J. Opondo. 1995. Land use change scenarios for subdivided ranches in *Laikipia* district, Kenya. *Laikipia*-Mt. Kenya paper No.19, LRP, Kenya.

Hunt-Davis Jr. 1986. Agriculture, Food, and the Colonial Period. In: Hansen, A. and McMillan, D.E (Eds). *Food in Sub-Saharan Africa*. Rynne Rienner, Boulder, Colorado, USA. pp 151-168

International Water Management Institute (IWMI). 2002. Changing the way we manage water to produce more food. CGIAR Challenge Program on Water for Food. *www.waterforfood.org*

IRIN. 2002. Kenya: Focus on Mt. Kenya water crisis. United Nations Integrated Regional Information Networks (IRIN). *http://www.irinnews.org/report.asp*

Isika, M., G.C.M. Mutiso and M. Muyanga. 2002. Kitui sand dams and food security. MVUA GHARP Newsletter. Vol. 4. April 2002 Edition. Kenya Rainwater Association, Nairobi.

Jaetzold, R. and H. Schidt. 1983. Farm management handbook of Kenya. Vol. IIB. Central Kenya Ministry of Agriculture, Nairobi

Jakeman, A.J. and R.A. Letcher. 2003. Integrated assessment and modelling: features, principles and examples for catchment management. *Environmental Modelling & Software, 18: 491-501*

Jodha, N.S. and A.C. Mascarenhas. 1985. Adjustment in Self-Provisioning Societies. *SCOPE* 27. John Wiley and Sons, New York, USA.

Kaumbutho, P.G. and J.M. Mutua. 2002. Experiences from on-farm conservation tillage trials in Kenya. Proceeding of a regional workshop on conservation tillage held at *Taj Pamodzi* Hotel, Lusaka, Zambia. 15-19 April, 2002. RELMA, Nairobi, Kenya. *63-110*

Kaumbutho, P.G. and F. Ochieng. 2001. The Kenya Conservation Tillage Initiative (KCTI): A report on KCTI conservation tillage trials for the short and long rains 2000. Kenya Network for Draught Animal Technology (KENDAT) and Regional Land Management Unit (RELMA), Nairobi, Kenya.

Khan, A.A.H. and C.K. Ong. 1997. Design and calibration of tipping bucket system for Field runoff and sediment quantification. *J. Soil and Water Conservation. 52(6): 437-443.*

Kiggundu, N. 2002. Evaluation of rainwater harvesting systems in Rakai and Mbarara Districts, Uganda. GHARP case study report. Greater Horn of Africa Rainwater Partnership (GHARP), Kenya Rainwater Association, Nairobi, Kenya.

Kiggundu, N. 1999.An economic analysis of a rainwater catchment system on a farmstead in *Sipili*, Kenya. Proc. of 9th International Rainwater Catchment

Systems Association (IRCSA), *Petrolina*, Brazil. July 6-9, 1999. *http://www.ircsa.org/9th.html*

Kihara, F.I. 2002. Evaluation of rainwater harvesting systems in *Laikipia* District, Kenya. GHARP case study report. Greater Horn of Africa Rainwater Partnership (GHARP), Kenya Rainwater Association, Nairobi, Kenya.

Kihara, F.I. and R. Ng'ethe. 1999. Water harvesting for enhanced crop production in *Laikipia* district, Kenya. Ministry of Agriculture, *Nanyuki*, Kenya.

Kijne. J. 2000. Water for food for Sub-Saharan Africa. Article prepared for FAO. *http://www.fao.org/landandwater/aglw/webpub/watfood.htm*

Kironchi, G. 1998. Influence of soil, climate and land use on soil water balance in the upper *Ewaso Ng'iro* basin in Kenya. Ph.D. thesis, Department of Soil Science, University of Nairobi, Kenya.

Kite, G., P. Droogers, H. Murray-Rust and K. de Voogt. 2001. Modelling scenarios for water allocation in the Gediz Basin, Turkey. IWMI Research Report No. 50. IWMI, Colombo, Sri Lanka. *p 29*

Kite, G.W. and P. Droogers. 2000. Integrated basin modelling. Research Report 43. IWMI, Colombo, Sri Lanka.

Kiteme, B.P. and J.K. Gikonyo. 2002. Preventing and resolving water use conflicts in the Mount Kenya highland-lowland system through water users' associations. *Mountain Research and Development, 22(1): 332-337*

Kithinji, G.R.M. and H.P. Liniger. 1991. Strategy for water conservation in *Laikipia* district. Proceedings of a water conservation seminar, Nanyuki, August 7-11 1991. *Laikipia*-Mt. Kenya papers, D-4. LRP and Universities of Nairobi (Kenya) and *Berne* (Switzerland).

Kohler, T. 1986 Land use and ownership. Institute of Geography, University of Berne, Switzerland.

Kutch, H. 1982. Principle features of a form of water concentrating culture. Trier Geographical Studies No. 5. Trier, Germany.

Lameck, P. 2002. Evaluation of rainwater harvesting systems in Dodoma District, Tanzania. GHARP case study report. Greater Horn of Africa Rainwater Partnership (GHARP), Kenya Rainwater Association, Nairobi, Kenya.

Leibundgut, C. 1986. Hydrogeological map of Mount Kenya area (1:50,000 map and exploratory text). *Geographica Bernensia, Vol. A3*, Berne, Switzerland.

LEISA. 1998. Challenging water scarcity. ILEIA Newsletter for the Low External Input and Sustainable Agriculture. 14(1). Editorial.

Liniger, H.P., J.K. Gikonyo, B.P. Kiteme and U. Wiesmann. 2005. Assessing and managing scarce tropical mountain water resources: The case of Mount Kenya and the semi-arid upper *Ewaso Ng'iro* basin. *Mountain Research and Development, 25(2): 163-173*

Liniger, H.P., C.N. Ondieki and G. Kironchi. 2000. Soil cover for improved productivity: Attractive water and soil conservation for the drylands in Kenya. *In:* Gichuki, F.N., D.N. Mungai, C.K.K. Gachene and D.B. Thomas *(eds)*. 2000. Land and water management in Kenya: Towards sustainable land use. Department of Agricultural Engineering, University of Nairobi, Nairobi, Kenya. *pp.* 223-231

Liniger, H.P., F.N. Gichuki, G. Kironchi, and L. Njeru. 1998. Pressure on the land: The search for sustainable use in a highly dense environment. *In:* Ojany, F.F *(ed.)*. 1998. Resource actors and policies: Towards sustainable regional development in the highland-lowland system of Mt. Kenya region. *Africa Geographical Journal. Vol. 8 (Special Edition)*.

Liniger, H.P. 1995. Endangered water: A global overview of degradation, conflicts and strategies for improvement. Development and Environment Report No. 12. Centre for Development and Environment, *Lang Druck* AG. Bern, Switzerland.

Liniger, H.P. 1991. Water conservation for rainfed farming in the semi-arid footzone northwest of Mt. Kenya (*Laikipia* highlands): Consequences on the water balance and soil productivity. *Laikipia*-Mt. Kenya Papers No. D-3. *Laikipia* Research Programme, Universities of Nairobi and *Bern*.

Lundgren, L.1993. Twenty years of soil conservation in Eastern Africa. Regional Soil Conservation Unit (RSCU) Report No. 9, Majestic Printing Works, Nairobi, Kenya. p 47

Mainuddin, M., S. Sangurunai, B. Kwanyuen and F. Penning de Vries. 2003. Management of land and water resources for sustainable small-holder Agriculture: Case studies in the *Mae Klong* river basin, Thailand. Final Report. International Water Management Institute (IWMI), Colombo, Sri Lanka.

Mathenge, G. 2002. Special Report: Laikipia District Constituency Profiles. Daily Nation, Monday, July 15, 2002. pp 11-15.

Mbugua, J. 1999. Rainwater harvesting and poverty alleviation: *Laikipia* experience. Proc. of 9th International Rainwater Catchment Systems Association (IRCSA), *Petrolina*, Brazil. July 6-9, 1999. *http://www.ircsa.org/9th.html*

Mbuvi, J. P and G. Kironchi. 1994. Explanations and profiles to reconnaissance soil survey of the upper *Ewaso Ng'iro* basin (*Laikipia* East and the slopes West to North of Mt. Kenya). *Laikipia*-Mt. Kenya Papers, B8

Meigh, J. 1995. The impact of small farm reservoirs on urban water supplies in Botswana. *Natural Resources Forum, 19 (1): 71-83*

Melesse, T., J. Rockström, G. Kidane and W. Berhe. 2002. Conservation tillage systems using improved implements for small-scale dryland farmers of Ethiopia. Proceeding of a regional workshop on conservation tillage held at *Taj Pamodzi* Hotel, Lusaka, Zambia. 15-19 April, 2002. RELMA, Nairobi, Kenya. *29-47*

Merrey, D.J., P. Drechsel, F.W.T. Penning de Vries and H. Sally. 2004. Integrating "livelihoods" into integrated water resources management: Taking the integration paradigm to its logical next step for developing countries. IWMI, Africa Regional Office.

Muni, R.K. 2002. Evaluation of rainwater harvesting systems in Machakos District, Kenya. GHARP case study report. Greater Horn of Africa Rainwater Partnership (GHARP), Kenya Rainwater Association, Nairobi, Kenya.

Mwakalia, S.S. and N. Hatibu. 1992. Rain water harvesting for crop production in Tanzania. Proc. 3rd Annual Conf. SADC, Land and water management research program. pp 513-525

Ngigi, S.N., H.H.G. Savenije, J.N. Thome, J. Rockström and F.W.T. Penning de Vries. 2005a. Agro-hydrological evaluation of on-farm rainwater storage systems for supplemental irrigation in *Laikipia* district, Kenya. *Agricultural Water Management, 73(1): 21-41*

Ngigi, S.N., H.H.G. Savenije, J. Rockström and C.K. Gachene. 2005b. Hydro-economic evaluation of rainwater harvesting and management technologies: Farmers' investment options and risks in semi-arid *Laikipia* district of Kenya. *Physics and Chemistry of the Earth, 30: 772-782*

Ngigi, S.N. 2003a. Rainwater harvesting for improved food security: Promising technologies in the Greater Horn of Africa. Greater Horn of Africa Rainwater Partnership (GHARP), Nairobi, Kenya. *p 266*

Ngigi, S.N. 2003b. What is the limit of up-scaling rainwater harvesting in a river basin? *Physics & Chemistry of the Earth, 28:943-956*

Ngigi, S.N. 2002a. Hydrological impacts of up-scaling rainwater harvesting on *Ewaso Ng'iro* river basin water management. Ph.D Research Proposal submitted to UNESCO-IHE, *Delft*, The Netherlands.

Ngigi, S.N. 2002b. Improved "dream" drip kits low-head irrigation technology in Kenya. 2nd International Contest for Innovative Irrigation Ideas and Technologies for Smallholders sponsored by the World Bank, Irrigation Association, IDE, IPTRID/FAO and Winrock International.

Ngigi, S.N. 2001. Rainwater harvesting supplemental irrigation: Promising technology for enhancing food security in semi-arid areas. Proceedings of the 10th IRCSA Conference. 10-14 September 2001, *Mannheim*, Germany.

Ngigi, S.N., J.N. Thome, D.W. Waweru and H.G. Blank. 2000. Low-cost irrigation for poverty reduction: An evaluation of low-head drip irrigation technologies in Kenya. IWMI Research Report. Collaborative research project between University of Nairobi and IWMI. IWMI Annual Report 2000-2001; pp. 23-29.

Ngigi, S.N. 1996. Design parameters for rainwater catchment systems in arid and semi-arid lands of Kenya. M.Sc. Thesis, Department of Agricultural Engineering, University of Nairobi.

Ngure, K.N. 2002. Evaluation of rainwater harvesting systems in Kitui District, Kenya. GHARP case study report. Greater Horn of Africa Rainwater Partnership (GHARP), Kenya Rainwater Association, Nairobi, Kenya.

Niederer, P. 2000. Classification and multi-temporal analysis of land use and land cover in the upper *Ewaso Ng'iro* basin (Kenya) using satellite and geographical information system (GIS). Unpublished M.Sc. thesis. Centre for Development and Environment (CDE), University of Berne, Berne, Switzerland.

Nietsch, S.L., J.G. Arnold, J.R. Kiniry and J.R. Williams. 2001. Soil and Water Assessment Tool (SWAT) User's Manual Version 2000. *http://www.brc.tamus.edu/swat/index.html*

NRM. 2003. Water abstractions monitoring campaign for the *Naro Moru* river, Upper *Ewaso Ng'iro* North Basin. Final Report. Natural Resources Monitoring, Modelling and Management (NRM), *Nanyuki*, Kenya

Oweis, T., A. Hachum and J. Kijne. 1999. Water harvesting and supplemental irrigation for improved water use efficiency. SWIM Paper No. 7. IWMI, Colombo, Sri Lanka. p 41

Pacey A. and A. Cullis. 1986. Rainwater harvesting: the collection of rainfall and runoff in rural areas. IT Publications, SRP, Exeter. London, UK.

Perrier, E.R. 1988. Water capture schemes for dryland farming. *In:* Unger, P.W., W.R. Jordan, T.V. Sneed and R.W. Jensen (*eds.*). 1988. Challenges in Dryland Agriculture-A Global Perspective. Proceeding of the International Conference on Dryland Farming. 235-238.

Presbitero, A.L. 2003. Soil erosion studies of humid Philippines. Ph.D Thesis, School of Environmental Studies, Griffith University, Queensland, Australia. *http://www4.gu.edu.au:8080/adt.root/public/adt-QGU20040909.151808/index.html*

Pwani, N.B. 2002. Personal communication. He is the Regional WatSan Program officer, International Federation of Red Cross and Red Crescent Societies, Regional Delegation Nairobi, Kenya.

Reij, C., I. Scoones and C. Toulin. 1996. Sustaining the soil: Indigenous soil and water conservation in Africa. EarthScan, London, UK. p 228

Reij, C., P. Mulder and L. Begemann. 1988. Water harvesting for plant production. World Bank Technical Paper No. 91. World Bank, New York, USA.

Rockström, J. 2001. Green water security for the food makers of tomorrow: Windows of opportunity in drought-prone savannahs. *Water Science and Technology, 43(4): 71-78*

Rockström, J., J. Barron and P. Fox. 2001. Water productivity in rainfed agriculture: Challenges and opportunities for smallholder farmers in drought-prone tropical agro-systems. Paper presented at an IWMI Workshop. Colombo, Sri Lanka. November 12-14, 2001.

Rockström, J. and M. Falkenmark. 2000. Semi-arid crop production from a hydrological perspective: Gap between potential and actual yields. *Critical Review Plant Science, 19(4): 319-346*

Rockström, J. 2000a. Water resources management in smallholder farms in Eastern and Southern Africa: An overview. *Phys. Chem. Earth (B), 25(3): 275-283*

Rockström, J. 2000b. Water balance accounting for design and planning of RWHS for supplemental irrigation. Technical Pamphlet No. 2. RELMA Publication, Nairobi, Kenya.

Rockström, J. 1999. On-farm green water estimates as a tool for increased food production in water scarce regions. *Phys. Chem. Earth (B), 24(4): 375-383*

Rockström, J., A. Kitalyi and P. Mwalley. 1999. Conservation tillage and integrated land management: Field experiences and extension approaches. Paper presented at the ATNESA/SANAT International Workshop on Empowering Farmers through Animal Traction into the 21st Century. South Africa, September 20-24, 1999.

Roth, S. 1997. Land use classification of the upper *Ewaso Ng'iro* basin in Kenya by means of Landsat TM satellite data. Institute of Geography, University of Bern, Bern, Switzerland

Savenije, H.H.G. 2004. The importance of interception and why we should delete the term evapotranspiration from our vocabulary. *Hydrological Processes,* 18(8):1507-1511.

Savenije, H.H.G. 1999. The role of green water in food production in Sub-Saharan Africa. FAO. *http://www.fao.org/ag/agl/aglw/webpub/greenwat.htm*

Savenije, H.H.G. 1997. Determination of evaporation from a catchment water balance at a monthly time scale, *Hydrology and Earth System Sciences*, Vol 1, No. 1, pp.93-100.

Schulze, R.E. 2002. Hydrological modelling: Concepts and practice. Unpublished lecture notes, HH062/02/1. UNESCO-IHE, Delft, The Netherlands. pp 160

Shanan, L. and N.H. Tadmor, 1976. Micro-catchment systems for arid zone development. Hebrew University of Jerusalem and Ministry of Agriculture, Rehovot, Israel.

Sharma, T.C. 1994. Stochastic features of drought in Kenya, East Africa. *Stochastic and Statistical Methods in Hydrology and Environmental Engineering, Vol.1: 125-137.*

Sharma, T.C. 1993. A Markov model for critical dry and wet days in *Kibwezi*, Kenya. *In:* Gichuki; F.N., D.N Mungai, C.K.K. Gachene and D.B. Thomas (*eds*.).1993. Proc. 4th Land and Water Management in Kenya: Towards Sustainable Land Use. Department of Agricultural Engineering, University of Nairobi: 233-237.

Sivapalan, M., K. Takeuchi, S.W. Franks, V.K. Gupta, H. Karambiri, V. Lakshmi, X. Liang, J.J. McDonnell, E.M. Mendiondo, P.E. O'Connell, T. Oki, J.W. Pomeroy, D. Schertzer, s. Uhlenbrook and E. Zehe. 2003. IAHS decade on Predictions in Ungauged Basins (PUB) 2003-2012: Shaping an exciting future of hydrological sciences. *Hydrological Sciences, 48(6), 857-880*

SIWI. 2001. Water harvesting for upgrading rainfed agriculture. Problem analysis and research needs. SIWI Report No. 11. Stockholm International Water Institute (SIWI), Stockholm, Sweden. p 97

Sombroek W.G, M.H. Brun and B.J.A. Van Der Pouw. 1980. The explanatory soil map and agro-climatic zone map of Kenya report No. E1, Kenya Soil Survey, Nairobi.

Thomas, D.B. (ed.). 1997. Soil and water conservation manual for Kenya. SWCB, Ministry of Agriculture and Livestock Development and Marketing. English Press, Nairobi, Kenya. p 294

Thomas, M.K., S.G. Mbui, G.G. Mikwa, P.W. Kung'u, G.K. Kiragu, J.N. Njeru, B. Gitari, L.C. MacMillian and J.K. Gikonyo. 1996. Flood utilization study. A report compiled for Ministry of land Reclamation, Regional and Water Development (*Laikipia* District) and Natural Resources Monitoring, Modelling and Management (NRM), *Nanyuki*, Kenya

Thome, J.N. 2005. Evaluation and design of runoff storage systems for supplemental irrigation in *Laikipia* District, Kenya. Unpublished M.Sc. Thesis. Department of Environmental and Biosystems Engineering, University of Nairobi, Kenya

Thurlow, T.L. and D.J. Herlock. 1993. Range management handbook of Kenya. Republic of Kenya, Ministry of Agriculture, Livestock Development and Marketing, Nairobi.

Tiffen, M., M. Mortimore, and F. Gichuki. 1994. More people less erosion: environmental recovery in Kenya. ACTS Press, African Centre for Technology Studies, Nairobi, Kenya. p 301

UNESCO. 2005. **H**ydrology for the **E**nvironment, **L**ife and **P**olicy (HELP) program of the United Nations Educational, Scientific and Cultural Organization (UNESCO). *www.portal.unesco.org/sc_nat/ev.php*

UNESCO and WMO. 2001. The design and implementation strategy of the **H**ydrology for the **E**nvironment, **L**ife and **P**olicy (HELP) program. Doc. No. H00/1 produced by a HELP Taskforce. *www.unesco.org/water/ihp/help*

UNDP/UNSO. 1997. Aridity zones and dryland populations: An assessment of population levels in the world's drylands with particular reference to Africa. UNDP Office to Combat Desertification and Drought (UNSO), New York, USA.

Vertessy, R.A., T.J. Hatton, R.G. Benyon and W.R. Dawes. 1996. Long-term growth and water balance predictions for a mountain ash (*eucalyptus regnans*) forest catchment subject to clearing-felling and regeneration. *Tree Physiology, 16(1/2): 221-232*

Vincent, L.F. 2003. Towards a smallholder hydrology for equitable and sustainable water management. *Natural Resources Forum, 27: 108-116*

Wairagu, M.M. 2000. Rainfall-runoff relationships on the *Njemps* flats of *Baringo* district: Implications for water harvesting. *In:* Gichuki, F.N., D.N. Mungai, C.K.K. Gachene and D.B. Thomas (*eds*). 2000. Land and water management in Kenya: Towards sustainable land use. Department of Agricultural Engineering, University of Nairobi, Nairobi, Kenya. *pp. 253-256*

Weismann, U., F.N. Gichuki, B.P. Kiteme and H.P. Liniger. 2000. Mitigating conflicts over water resources in the highland-lowland system of Mount Kenya. *Mountain Research and Development, 20(1): 10-15*

World Bank. 1997. World Development Report. World Bank, Washington D.C, USA.

WRAP. 1987. Water Resources Assessment study in *Laikipia* district. Water Resources Assessment Program (WRAP), Ministry of Water Development, Nairobi, Kenya

Zöbisch, M.A., P. Klingspor, A.R. Oduor and R. Schlott. 2000. Laboratory calibration of tipping bucket device and sediment sampling tube for soil erosion plots. *In:* Gichuki; F.N., D.N Mungai, C.K.K. Gachene and D.B. Thomas (*eds.*).1993. Proc. 4th Land and Water Management in Kenya: Towards Sustainable Land Use. Department of Agricultural Engineering, University of Nairobi: *81-84*.

Annexes

Annex 1: Computation of supplemental irrigation requirement for 90-day growing period cabbage in *Matanya* for the short rains 2002

Crop stage	Date (10-day)	E_o (mm) [1]	$0.5E_o$ (mm) [2]	K_c^{14} [3]	T_c (mm) [4]	D_{rz} (mm) [5]	P (mm) [6]	P_e (mm) [7]	S_d (mm) [8]	S_s (mm) [9]	S_t (mm) [10]	S_{t-1} (mm) [11]	Q_{dp} (mm) [12]	Q_i (mm) [13]
Initial	19-28/10/02	61.2	24.5	0.45	22.0	100	10.0	41.6	11.5	31.0	60.0	79.6	0.0	0.0
	29/10-7/11/02	36.4	14.6	0.45	13.1	100	131.4	99.6	5.6	92.1	81.7	168.2	33.7	0.0
vegetative	8-17/11/02	49.5	19.8	0.75	29.7	200	35.5	26.0	18.7	15.0	176.9	173.2	0.0	0.0
	18-27/11/02	49.5	19.8	0.89	35.7	400	6.0	4.0	33.6	1.9	72.3	57.4	0.0	16.8
	28/11-7/12/02	63.1	25.2	1.03	52.0	600	18.1	12.9	39.1	0.0	0.0	0.0	0.0	39.1
Yield formation	8-17/12/02	42.8	17.1	1.03	35.3	600	55.8	40.6	21.9	27.2	77.6	98.7	0.0	15.7
	18-27/12/02	58.9	23.6	1.03	48.5	600	47.1	32.5	20.3	4.3	90.2	74.1	0.0	0.0
	28/12/20 - 6/1/2003	42.8	17.1	1.03	35.3	600	24.7	18.2	28.1	11.0	37.7	32.8	0.0	12.2
Maturity	7-16/1/03	56.0	22.4	0.95	42.6	600	0.0	0.0	42.6	0.0	0.0	0.0	0.0	42.6

Note:

☞ The units are in mm 10-day^{-1} (i.e. per a 10 day period).

☞ [2] = 0.5*0.8*[1]; [4] = [2]*[3]; [7] = IF([6]<2.3, 0.6*[6]-0.3, 0.8*[6]-0.8); [8] = IF([4]-[7]<0, [4]-[7]); [9] = IF([7]-[4]<0, 0, [7]-[4]); [10] = IF([11]<0, 0, IF([11]>S_{max}, S_{max}, [11]))^{15}; [11] = IF([10]-[8]+[9]<0, 0, [10]-[8]+[9]); [12] = IF([11]> S_{max}, [11]-S_{max}, 0); [13] = IF([10]<[8], [8]-[10], 0)

^{14} K_c = 0.45 for first 20 days (initial stage), 0.75 for the next 15 days (vegetative stage), 1.03 for the next 45 days (yield formation) and 0.95 for the last 10 days (maturity).

^{15} S_{max} = 27mm at D_{rz} = 100mm; S_{max} = 69mm at D_{rz} = 200mm; and S_{max} = 250mm at D_{rz} = 600mm.

Annex 2: Sample computation of additional soil moisture (θ_s) due to conservation tillage on farmers' fields for four seasons (2002-4)

Farmer's field	Average Y_{TT} (kg ha^{-1})	Average Y_{CT} (kg ha^{-1})	Yield increase (%)	Soil moisture increase (%)
1	3,200	3,827	20	15
2	1,392	1,806	30	22
3	3,281	3,958	21	16
4	2,854	4,272	50	36
5	3,548	4,289	21	16
6	4,400	5,300	20	16
7	3,286	4,086	24	19
8	4,444	6,528	47	34
9	3,910	4,674	20	15
10	8,25	9,98	21	16
11	2,800	3,950	41	30
12	2,840	3,398	20	15
13	2,548	3,273	28	22
14	1,765	2,234	27	20
15	4,012	5,285	32	24
16	3,643	4,556	25	19
Average	3,047	3,902	28	21

Annex 3: Sample runoff data and computation of runoff reduction (%) due to conservation tillage on a runoff plot and a farmer's field

Rainfall (mm day⁻¹)	Runoff plot			Farmer's field		
	TT plot runoff (mm day⁻¹)	CT plot runoff (mm day⁻¹)	Runoff reduction (%)	TT field runoff (mm day⁻¹)	CT field runoff (mm day⁻¹)	Runoff reduction (%)
22.4	8.2	5.1	60	5.4	3.8	29
26.6	10.3	7.6	36	8.2	6.0	27
45.6	19.8	16.1	23	9.8	7.5	24
56.8	25.4	20.6	23	5.8	4.2	28
16.8	5.4	4.5	20	4.2	3.2	24
24.6	9.3	7.6	23	4.4	2.6	41
29.0	11.5	9.6	20	7.9	5.7	27
27.6	10.8	8.6	26	4.9	3.8	22
26.6	10.3	7.3	41	4.0	2.6	35
34.0	14.0	11.5	21	6.4	5.4	16
42.0	18.0	14.4	25	4.3	3.4	21
30.6	12.3	8.0	54	20.2	14.8	27
24.0	9.0	6.2	46	5.6	4.4	21
17.6	5.8	4.6	26	4.8	4.0	18
36.2	15.1	11.0	37	10.6	8.4	21
12.4	3.2	2.6	23	16.4	10.7	35
9.4	1.7	1.4	21	3.8	2.3	38
32.0	13.0	10.8	21	5.1	4.0	22
23.6	8.8	6.5	35	7.3	4.8	34
15.6	4.8	3.8	25	5.6	4.5	20

Note: The sample runoff data are for a selected runoff plot and farmer's field during four rainy seasons (2002–4) in *Kalalu, Laikipia* district, Kenya.

Annex 4: Computation of impacts of 100% flood storage on dry season irrigation water abstraction and downstream river flows at A6

Month	A6 measured flow, m³s⁻¹ [1]	Naturalized flows, m³s⁻¹ [2]	Irrigation demand, Mm³/month [3]	Irrigation water supply, Mm³/month		Release, Mm³/month [6]	Balance, Mm³ [7]	Spill, Mm³/month [8]	Storage, Mm³ [9]	Downstream flow, m³s⁻¹ [10]	Water resources, m³s⁻¹ [11]
				Storage [4]	Abstraction [5]						
Apr.	1.25	1.76	0.0	0.0	0.0	0.0	4.57	0.00	4.57	0.00	0.00
May	1.85	2.65	0.0	0.0	0.0	0.0	11.44	3.67	7.77	1.42	1.42
Jun.	0.24	0.75	1.6	1.6	0.0	1.95	6.22	0.00	6.22	0.75	1.35
Jul.	0.08	0.42	1.6	1.6	0.0	1.08	4.66	0.00	4.66	0.42	1.02
Aug.	0.10	0.44	1.6	1.6	0.0	1.13	3.11	0.00	3.11	0.44	1.04
Sep.	0.06	0.40	1.6	1.6	0.0	1.03	1.55	0.00	1.55	0.40	1.00
Oct.	0.18	0.69	1.6	1.6	0.0	1.80	0.00	0.00	0.00	0.69	1.29
Nov.	1.64	2.44	0.0	0.0	0.0	0.0	6.33	0.00	6.33	0.00	0.00
Dec.	0.92	1.43	0.0	0.0	0.0	0.0	10.04	2.27	7.77	0.88	0.88
Jan.	1.00	1.51	2.6	2.6	0.0	3.91	5.18	0.00	5.18	1.51	2.51
Feb.	0.13	0.46	2.6	2.6	0.0	1.20	2.59	0.00	2.59	0.46	1.46
Mar.	0.32	0.83	2.6	2.6	0.0	2.16	0.00	0.00	0.00	0.83	1.83
Mean	0.65	1.15								0.65	1.15

Note:

☞ Irrigable area = 1,500 ha; Irrigation efficiency = 50%; Storage efficiency = 100% and Maximum storage = 7.8 Mm³ for 3 months

☞ $[6] = [2]*2.59 - [5]$; $[7] = [9]_{k-1} + 2.59*[2] - ([4] + [5] + [6])$; $[8] = Max.([7] - Max., 0)$; and $[10] = ([6] + [8])/2.59$

Annex 5: Computation of impacts of 75% flood storage on dry season irrigation water abstraction and downstream river flows at A6

Month	A6 measured flow, m³s⁻¹ [1]	Naturalized flows, m³s⁻¹ [2]	Irrigation demand, Mm³/month [3]	Irrigation water supply, Mm³/month		Release, Mm³/month [6]	Balance, Mm³ [7]	Spill, Mm³/month [8]	Storage, Mm³ [9]	Downstream flow, m³s⁻¹ [10]	Water resources, m³s⁻¹ [11]
				Storage [4]	Abstraction [5]						
Apr.	1.25	1.76	0.0	0.0	0.0	0.44	4.13	0.00	4.13	0.17	0.17
May	1.85	2.65	0.0	0.0	0.0	0.66	10.34	4.51	5.83	2.00	2.00
Jun.	0.24	0.75	1.6	1.2	0.4	1.56	4.66	0.00	4.66	0.60	1.20
Jul.	0.08	0.42	1.6	1.2	0.4	0.69	3.50	0.00	3.50	0.27	0.87
Aug.	0.10	0.44	1.6	1.2	0.4	0.75	2.33	0.00	2.33	0.29	0.89
Sep.	0.06	0.40	1.6	1.2	0.4	0.64	1.17	0.00	1.17	0.25	0.85
Oct.	0.18	0.69	1.6	1.2	0.4	1.41	0.00	0.00	0.00	0.54	1.14
Nov.	1.64	2.44	0.0	0.0	0.0	0.61	5.72	0.00	5.72	0.24	0.24
Dec.	0.92	1.43	0.0	0.0	0.0	0.36	9.07	3.25	5.83	1.39	1.39
Jan.	1.00	1.51	2.6	1.9	0.6	3.27	3.89	0.00	3.89	1.26	2.26
Feb.	0.13	0.46	2.6	1.9	0.6	0.55	1.94	0.00	1.94	0.21	1.21
Mar.	0.32	0.83	2.6	1.9	0.6	1.51	0.00	0.00	0.00	0.58	1.58
Mean	0.65	1.15								0.65	1.15

Note:

☞ Irrigable area = 1,500 ha; Irrigation efficiency = 50%; Storage efficiency = 100% and Maximum storage = 5.8 Mm³ for 3 months

☞ $[6] = [2]*2.59 - [5]; [7] = [9]_{t-1} + 2.59*[2] - ([4] + [5] + [6]); [8] = Max.([7] - Max., 0);$ and $[10] = ([6] + [8])/2.59$

Hydrological Impacts of Land Use Changes

Annex 6: Computation of impacts of 50% flood storage on dry season irrigation water abstraction and downstream river flows at A6

Month	A6 measured flow, m³s⁻¹ [1]	Naturalized flows, m³s⁻¹ [2]	Irrigation demand, Mm³/month [3]	Irrigation water supply, Mm³/month		Release, Mm³/month [6]	Balance, Mm³ [7]	Spill, Mm³/month [8]	Storage, Mm³ [9]	Downstream flow, m³s⁻¹ [10]	Water resources, m³s⁻¹ [11]
				Storage [4]	Abstraction [5]						
Apr.	1.25	1.76	0.0	0.0	0.0	0.88	3.69	0.00	3.69	0.34	0.34
May	1.85	2.65	0.0	0.0	0.0	1.33	9.23	5.35	3.89	2.58	2.58
Jun.	0.24	0.75	1.6	0.8	0.8	1.17	3.11	0.00	3.11	0.45	1.05
Jul.	0.08	0.42	1.6	0.8	0.8	0.31	2.33	0.00	2.33	0.12	0.72
Aug.	0.10	0.44	1.6	0.8	0.8	0.36	1.55	0.00	1.55	0.14	0.74
Sep.	0.06	0.40	1.6	0.8	0.8	0.25	0.78	0.00	0.78	0.10	0.70
Oct.	0.18	0.69	1.6	0.8	0.8	1.02	0.00	0.00	0.00	0.39	0.99
Nov.	1.64	2.44	0.0	0.0	0.0	1.22	5.11	1.22	3.89	0.94	0.94
Dec.	0.92	1.43	0.0	0.0	0.0	0.72	6.88	3.00	3.89	1.43	1.43
Jan.	1.00	1.51	2.6	1.3	1.3	2.62	2.59	0.00	2.59	1.01	2.01
Feb.	0.13	0.46	2.6	1.3	1.3	-0.10	1.30	0.00	1.30	-0.04	0.96
Mar.	0.32	0.83	2.6	1.3	1.3	0.86	0.00	0.00	0.00	0.33	1.33
Mean	0.65	1.15								0.65	1.15

Note:

☞ Irrigable area = 1,500 ha; Irrigation efficiency = 50%; Storage efficiency = 100% and Maximum storage = 3.9 Mm³ for 3 months

☞ [6] = [2]*2.59 − [5]; [7] = [9]$_{I-1}$ + 2.59*[2] − ([4] + [5] + [6]); [8] = Max.([7] − Max., 0); and [10] = ([6] + [8])/2.59

About the Author

 Stephen N. Ngigi was born in *Nyandarua*, Kenya on December 17, 1967. He obtained his B.Sc. (*Hons.*) in Agricultural Engineering from the University of Nairobi in 1992. He worked in the same University as a Graduate Assistant before proceeding for a diploma in groundwater resources management at the Hebrew University of Jerusalem (Israel) in 1993. He obtained his M.Sc. in Agricultural Engineering (*Soil and Water Engineering Option*) from the University of Nairobi in 1996. He then worked for an international NGO (Africa Centre for Technology Studies (ACTS)) before rejoining the University of Nairobi as a Tutorial Fellow in 1998. This was a training position that entailed research and development of a Ph.D proposal. In 2001, he successfully applied for WOTRO fellowship through ILRI, a CGIAR centre in Nairobi. The fellowship could only cover part of the research budget, but IWMI accepted to co-fund his Ph.D study. After securing Ph.D scholarship, he was promoted to a Lecturer position and immediately applied for a study leave and joined UNESCO-IHE as a promovendus in May 2002.

His interest in RHM dates back in 1993 during the 6th IRSCA conference held in Nairobi. His M.Sc. research on "*Design parameters for rainwater catchment systems in arid and semi-arid lands of Kenya*" was his first undertaking. In his belief that RHM is one of the solutions for improving food security in drought prone SSA, he developed and implemented a USAID-funded regional project that evaluated the performances and shortcomings of some promising RHM systems. He published the results in a book entitled "*Rainwater harvesting for improved food security: Promising technologies in the Greater Horn of Africa*". The results also formed the background of his Ph.D research. As a follow up to the evaluation project, he developed and led implementation of three pilot projects, funded by USAID and DED, to demonstrate the viability of some of the promising RHM systems to improve livelihoods in semi-arid environment. He has also been involved in promoting RHM systems in the *Nuba* Mountains, southern Sudan

He has also been involved in a number of IWMI's regional projects in Kenya, which culminated into a publication entitled "*The changing face of irrigation in Kenya: Opportunities for anticipating change in eastern and southern Africa*" in which he contributed two chapters. He was also involved in research and development of a locally designed low-head drip irrigation systems in Kenya. This was conceived out of a challenge by smallholder farmers who were experiencing problems with the World Bank supported project, which was promoting USA based *Chapin Watermatics* drip systems. The initiative was also supported by IWMI. The innovation was entered to the international contest for innovative irrigation technologies for smallholders in 2001 and 2002. He was ranked 3rd and 2nd runners up respectively among many contestants from all over the world.

He has taught and researched on water resources development and management related topics since he joined the University. He has also presented many papers in international conferences and a number of his papers have been published in international journals. He hopes to influence policy to improve food security, water resources management and livelihoods by upgrading rainfed agriculture through adoption of viable RHM systems in SASE of SSA.

Printed and bound by CPI Group (UK) Ltd, Croydon, CR0 4YY

22/10/2024

01777530-0006